Left to Our Own Devices

Left to Our Own Devices

Outsmarting Smart Technology to Reclaim Our Relationships, Health, and Focus

Margaret E. Morris

Foreword by Sherry Turkle

The MIT Press
Cambridge, Massachusetts
London, England

This book was set in Stone Serif by Westchester Publishing Services.

Library of Congress Cataloging-in-Publication Data
Names: Morris, Margaret E., author.
Title: Left to our own devices : outsmarting smart technology to reclaim our
   relationships, health, and focus / Margaret E. Morris ; foreword by Sherry Turkle.
Description: Cambridge, MA : MIT Press, [2018] | Includes bibliographical
   references and index.
Identifiers: LCCN 2018013341 | ISBN 9780262039130 (hardcover : alk. paper)
ISBN 9780262552066 (paperback)
Subjects: LCSH: Internet in psychotherapy. | Internet--Psychological aspects.
   | Internet--Social aspects. | Health. | Interpersonal relations.
Classification: LCC RC489.I54 M67 2018 | DDC 616.89/14--dc23 LC record
   available at https://lccn.loc.gov/2018013341

To my brother Michael, who has always pushed me to move forward

Contents

Foreword

Sherry Turkle

Margaret Morris and I would seem to be on opposite sides of an argument. I am a partisan of conversation. Morris is a maestro of apps. Readers of this book will learn that this is too simple a story. In real life, when you take the time to look closely, we all talk back to technology. Or want to. The most humane technology makes that easy.

That puts a responsibility on designers. And on those of us who bring technology into our everyday lives. To make more humane technology, we have to make it our own in our own way. We can't divide the world into builders and users. Digital culture needs participants, citizens.

So if my plea to those who would build "empathy apps" has always been, "We, people, present, talking with each other, we are the empathy app," both Morris and I would ask, "Well, how can we build technologies that encourage that conversation?"

For many years, I have written about the power of *evocative objects* to provoke self-reflection. But some objects, and by extension, some technologies, are more evocative than others. *Left to Our Own Devices* can be read as a primer for considering what might make for the most evocative technology. And if you suspect you have one in your hand, how might you best use it? You can reframe this question: If you are working with a technology that might close down important conversations, can it be repurposed to open them up?

Indeed, my first encounter with Morris was in June 2005, when she wrote me about a technology that I was already worried about.

The object in question was the robot, Paro, a sociable robot in the shape of a baby seal, designed to be a companion for the elderly. Paro gives the impression of understanding simple expressions of language and emotion.

It recognizes sadness and joy, and makes sounds and gestures that seem emotionally appropriate in return. Its inventor, Takanori Shibata, saw Paro as the perfect companion for the elderly, and so did a lot of other people. After I visited Japan in fall 2004, Shibata gave me three Paros, and I began to work with them in eldercare facilities in Massachusetts. That's what I was doing when I received Morris's first email. She said that she was a clinical psychologist interested in the "health benefits" of Paro. Shibata suggested that she be in touch with me. I look at that December 2005 email now and note my polite response: "Yes, surely, let's talk."

I dreaded the conversation with Morris. I didn't have cheerful things to say to someone working on robots, health apps, and health benefits. The Paro project troubled me. Robots like Paro could pick up on language and tone, and offer pretend empathy to the elderly. But the robot understood nothing of what was said to it. Every day when I went to work, I was asking myself, "What is pretend empathy good for?"

One day my conflicts came to a head: an old woman whose son had just died shared the story with Paro, who responded with a "sad" head roll and sound. The woman felt understood. I was there with my graduate students and a group of nursing home attendants. Their mood was celebratory. *We had gotten the woman to talk to a robot about something important* and somehow that seemed a success. Yet in that so celebrated exchange, *the robot was not listening to the woman*, and we, the researchers, were standing around, watching. We, who could have been there for her and empathized, were happy to be on the sidelines, cheering on a future where pretend empathy would be the new normal. I was distressed.

So I approached my conversation with Morris thinking that perhaps I could be in dialogue with her as a kind of respectful opposition. But as soon as we met, it became clear that this would not be our relationship at all. Our conversations about Paro were not about any simple notion of benefits. She understood my concerns, and we moved to this: Were there examples from my research where robots had opened up a dialogue? How could the presence of humans with the robots help this happen? What are the situations where a human-technology relationship is mediated by a human-human relationship that brings people closer to their human truth? In the end, the pretend empathy of sociable robots is not a place where I was or am comfortable, but these questions were right.

And Morris's own work brought them to the foreground. I remember the first of her case studies that she discussed with me, where her presence had changed how a family experienced a social activity tracker. (The story she shared appears here as "Family Planets" in chapter 3.) An older woman was using the tracker. Her concerned daughter was in on the results. On the surface, the family's use of the tracker was instrumental: Could feedback from the social activity tracker convince the mother that more social stimulation was good for her?

In practice, the technology became a bridging device to open up a conversation between the daughter and her mother that the daughter did not know how to start alone. Now, with the app, the daughter found new words. The app had a display that showed her mother as a circle, an island that did not intersect with others. Now the daughter spoke of her mother's isolation as "like being on an island, when everyone else you've known and loved has died." These were thoughts that the technology gave her permission to articulate. And Morris's presence, too, gave the daughter courage. Through the tracker, the daughter, and Morris, the daughter and the mother formed a bond; through the tracker, the daughter felt empowered because she had a scientist on her side. Morris helped the daughter interpret an independent view of her mother's isolation. It was nothing to be ashamed of. It was not an accusation. But it was not healthful, and people were here to help.

Here, the tracking technology was an evocative object for talking about feelings, an externalization, and its snapshot of the inner life facilitated new conversations. *People are the empathy app, but technology can help them get more comfortable in that role.* In 2005, Morris and I began a relationship that has never been about my being the opposition, even when we disagree. It is simply about conversation. Which technology opens it up? Which technology closes it down? The fact that a technology is evocative only means it has a potential that can cut both ways. Its holding power can be used to compel you to waste your time on social media that breaks down your sense of autonomy and pride in your own accomplishments. Or it can suggest new framings and new ways of thinking.

Which way it goes can be inflected by the design of the technology. Yet equally critical is the culture that is built around that technology. In the world that we need to build with technology, we don't need critics and enthusiasts. We need to wear down the wall between "users" and designers.

For me, sharing responsibility for technology is the path toward a humane technology that supports us in the lives we want to live.

Morris encourages this kind of thinking because she's spent her professional lifetime paying attention to how people make technology their own in their own way. The stories in this book are the *real* conversations that happen between people and objects. They are almost always not the conversations that were imagined by the designers. They are surprising. Sad. Funny. Hopeful. Human.

Acknowledgments

I have many to thank for helping me complete this book. I am grateful, first, to the people I've interviewed over the last fifteen years. Some of their stories are recounted here. These individuals generously shared their experiences with me and often allowed me into their homes, routines, and relationships. They educated me about their struggles, from regulating blood sugar to coping with the loss of a significant other, often over a series of conversations.

I am also grateful to the researchers and practitioners who have described their work to me. I owe particular thanks to Nazanin Andalibi, Alison Darcy, Anna Wexler, Natasha Dow Schüll, Jessica Floeh, Ronni Higger, Elizabeth Bales, Jennifer Veitch, Mark Griswold, Jonathan Wright, Craig Mundie, Kane Race, Kristina Olson, Oliver Haimson, Christina Chung, Lauren McCarthy, Pamela Hinds, Wendy Ju, Stephen Schueller, and Stephanie Zerwas.

I am grateful to the many mentors and helpful collaborators I have had while conducting the research described in this book. While working at Intel, I had the benefit of learning from Eric Dishman and Genevieve Bell, two social scientists who made a huge mark on the company and created a space for me to explore how computing could foster intimacy and health. Maria Bezaitis, my manager and mentor, patiently read my early proposals for this book. Many others at Intel contributed to the projects I describe or my thinking about them. In particular, I am indebted to Farzin Guilak, Melissa Greg, John Sherry, Heather Patterson, Bill DeLeeuw, Qusai Kathawala, Jay Lundell, Tim Brooke, Stefanie Danhope, and Emma Shepanek, my collaborator for the research on online dating, and the Immersive Computing Lab. My academic collaborations, especially those with Luis Ceze, Stephen Intille, Richard Sloan, Ethan Gorenstein, Misha Pavel, Gillian Hayes,

Sean Munson, Conrad Nied, and Gary Hsieh, made many of these projects possible.

I thank Doug Carmean, who started out as my best collaborator and became my life partner. Doug is a renowned computer architect and also an artist who creatively brought psychological concepts to life with immersive computing, communication platforms, and connected devices. He developed a number of the systems described here. My thinking has been shaped by his astute observations of how people used those systems and the interactions he notices in daily life. This book has long been the center of our living space and conversations—one of many reasons I am grateful to him.

This book was refined as I taught research groups in the department of Human Centered Design and Engineering at the University of Washington. I am grateful to David McDonald for this opportunity and Daniela Rosner for advising me on design exercises. Teaching allowed me to integrate ideas related to identity and appropriation, and observe students grappling with these ideas as they reflected on their own use of technology. I thank the students in these classes who commented on some of the examples in the book. In particular, I thank Andrea Sequeira for her insightful review.

I am grateful to Gita Manaktala, editorial director at the MIT Press, for guiding me and my manuscript from the seeds of a proposal through publication. Along the way, I was extremely fortunate to work with editors Ada Brunstein and David Weinberger, who helped me focus on the key theme of taking ownership of technology to enhance relationships and health. I also appreciate the thoughtful copyediting of Deborah Cantor-Adams and Sandy Kaplan. Crystal Rutland, my close friend and colleague, critiqued my text and encouraged me in ways that were essential. I thank danah boyd for her words of wisdom about finishing a book. In various stages, this manuscript benefited from the reviews of anonymous peers and three great minds: Elizabeth Churchill, Natasha Dow Schüll, and Gary Wolf.

As I started this project years ago, I was fortunate to work with James Levine, who encouraged me to organize the book around a set of stories. I followed that advice. The stories have grown and have more discussion around them than they once did, but they remain the core of this book.

This book is inspired in large part by an extended conversation with MIT professor and author Sherry Turkle starting in 2005. At that point, I had shifted gears professionally, from clinical psychology to developing ideas for technologies to promote psychological well-being. Professor Turkle, of course, was and is the most prominent scholar on our relationships with technology and its influence on psychological development. Despite her growing concern about the risks of technology, she supported my attempts to develop systems that could be psychologically helpful. Our conversations about how people used the early prototypes that my colleagues and I developed for connectedness and emotional regulation sparked an insight that has guided my work since: it was the conversations and human interactions evoked by technologies, not the systems themselves, that made a difference in people's lives. In my exchanges with Professor Turkle and in her writing, I found my intellectual lineage and my calling. I could see how my psychological training, as a listener and interpreter, although not put into therapy practice as I had originally envisioned, could contribute to the world. Given that Professor Turkle inspired this journey, it is a great honor to have her words open this book.

I've spent almost two decades talking with people about what they want to change in their lives and how that ties in with their devices. I've looked for bright spots, where people found ways to use technology constructively, often in what I call "off-label uses" that the technology designer never intended. It's my hope that readers will be inspired by these stories to take ownership of their devices as they reach toward connectedness and health.

Introduction

"How could a phone be a shrink?" This question drove my research at Intel in 2006 and led to a prototype we called the Mood Phone. I knew that the idea flew in the face of an implicit tenet of therapy: that unmediated interpersonal dialogue was essential for it to be effective. But I also knew that the traditional model of therapy was itself constrained by the technological limitations of the age in which it first developed.

I had spent the better part of a decade training to become a clinical psychologist. It was, I felt, my calling. I had seen how powerful individual therapy could be, and how deeply it could help people emotionally and in their relationships. But I also knew that the "talking cure"—the late nineteenth-century paradigm of an extended dialogue between a therapist and client held in a setting removed from the client's everyday life—didn't scale. A good therapist is expensive, physically distant, and available by appointment only. Our problems occur in the mix of our lives, unscheduled.

Therapists have traditionally been unavailable to their clients between sessions. Historically, therapy has been based on retrospective self-report, not because that is sufficient or accurate, but because there wasn't an obvious way to capture a patient's experiences in context. And therapy has been confined to the therapist and client, not because it wouldn't be helpful to have consultation from additional experts or peers, but because it wasn't feasible to incorporate such therapeutic extensions into the dialogue. These and other limitations have been embraced and justified as principles of the practice, yet they can also be seen as hurdles that were insurmountable with the technology of the time.[1]

The internet, and then the near ubiquity of mobile phones by the early 2000s, enabled new types of communication media that let us reenvision therapy. We, as psychologists and individuals concerned about our relationships, were cautioned that these new media diminished communications and trust, distorted feelings and meaning, and were giving rise to new disorders and even a new type of addiction.

I was unconvinced.

So I began the Mood Phone effort as an experiment with colleagues at Intel and Columbia University. The Mood Phone was an app designed to serve as a personal therapeutic agent, whose interactive prompts were rooted in the psychological principles used by therapists, specifically cognitive behavioral therapy. We wanted to offer individuals a digital therapist at their disposal. The Mood Phone integrated sensors, calendar prompts, and self-tracking data to detect emotional changes. It offered visual and verbal cues to help individuals navigate their problems in real time. These cues were inspired primarily by cognitive behavioral therapy, but also by yoga and mindfulness. We involved participants from the earliest iterations. We gave them prototypes to use in their daily lives. We wanted to see *how* they would use this on-the-spot therapy, not just whether a final prototype was usable.

One of the early hints that this tool would be more than a private therapist in your pocket came from a participant, Chandra, who told me how she used it to manage a toxic interaction.[2] She had walked into a bar to meet friends when she heard one of them bad-mouth a mutual friend who wasn't there. It got uglier, turning into a character assault. Chandra held up the Mood Phone and confronted the accuser with a screen that asked rhetorically, "Might I be villainizing?" Villainizing, an extension of tendencies to see the world in black-and-white white terms, is part of an attributional style associated with hostility.[3] By holding a mirror up in this way, Chandra interrupted the attack on her friend and perhaps even encouraged some self-reflection among her friends in the bar.

Chandra took the Mood Phone app beyond its intended use, challenging the intent of the app and also the way that I had been trained to think about individual therapy. In cognitive behavioral therapy, clients learn to examine and reframe their own thoughts, not those of others. The Mood Phone had been designed with the assumption that it would be an always-handy extension of the therapist, available to help individuals address their own problems. We had not anticipated that people would use the phone to coach or confront others, as Chandra and many did. Of course, they also used it for self-reflection, but it was notable that applying the app socially did help many of these individuals manage their own stress.

Chandra had taken a technology carefully designed for one purpose and extended its use for another. She had turned it into a social tool appropriate for her situation. Perhaps much of the concern about communications technology came from an assumption that people would use these tools exactly in the way that the designers had anticipated. What if we looked at how people are making these tools their own, going beyond the expected uses? What if we looked more fully at how these tools are being used in the complex social situations of daily life?

My interviews throughout the Mood Phone project continued to impress on me how putting psychotherapeutic services on a mobile phone changed the therapy from a private discussion removed from the contexts of daily life to a form of dialogue that people brought into their routines and conversations. The sharing of the Mood Phone typically occurred in a more serious way and in closer relationships (with spouses, children, and close colleagues) than Chandra's playful confrontation in the bar. But in all these instances, the interpersonal use of this personal app seemed to enrich its value for the primary user and may have also helped those with whom they shared it.

In the years since that project, I have experimented with many other ways of bringing what psychologists know about emotion, communication, and health into the technologies that people use throughout the day. I have also watched the ways that individuals use popular products, such as ride sharing and dating apps. The lesson that I first learned from Chandra—that benefit often comes as people break or expand the rules to depart from the intended usage—has played out repeatedly. I have seen that our relationship to technology and the benefits we reap from it depend on how much we make it our own.

This has motivated me to contextualize the drumbeat we hear about the perils of technology, particularly social media: increased isolation, difficulty empathizing, and impaired conversational skills. Sherry Turkle's compelling TED talk about the isolating effects of technology has been viewed over four million times.[4] This talk resonates with a desire for more connectedness along with a growing concern about the distraction that we see in ourselves and others. It has been found, for example, that the majority of individuals in the United States who use a mobile phone have felt ignored because of how much time someone else in their household spends on a phone.[5] Negative feelings are especially likely to arise when someone's technology use violates the expectations set within a close relationship.[6] Despite these strong concerns, it doesn't seem like most people are ready to abandon their devices.

Most of us rely on communication technology, especially the messaging, social networking applications, email and voice services on our phones, for connectedness with family and friends. This connectedness, through whatever means it is sustained, is vital. I suspect most people reading this have suffered through periods of feeling isolated or know someone who has. Loneliness is common, particularly among those under eighteen and over sixty-five, and poses health risks (e.g., for dementia and heart disease) that are comparable to obesity and smoking.[7] Like other health concerns, loneliness may spread within social networks.[8] It is obviously not the case that all communication works against loneliness, that every glance at Facebook or every composed email cultivates feelings of belongingness or closeness. Nor is it clear that phones, social media, or the internet cause isolation. To the contrary, some research associates internet use of any sort with increased communication and social satisfaction.[9] For those who are extraverted, these channels offer additional contact, and for those who are socially anxious, texting and online communication lower the barriers to communicating.[10] Many teens and kids find friendship as well as acceptance through social media that is not available in their local communities. Online communities are often especially critical for teens who feel ostracized due to their gender identity and sexual orientation.[11]

While social sites like Facebook and Twitter are generally viewed as echo chambers, they also host conversations among people who might otherwise feel disconnected. The 2017 explosion of "Me Too," which rekindled a movement begun in the era before hashtags, connected individuals who may have previously felt alone in their experience of sexual harassment and prompted significant action against high-profile perpetrators.[12] And, occasionally, online conversations with strangers provoke a major change in beliefs and bold rejections of discrimination.[13] While many of the most notable uses of social media bridge geographic and cultural distances, some people are leveraging it to connect with neighbors.[14]

I broaden the definition of social media to include all the technologies we use to connect—whether that is with fleeting acquaintances, close friends and family, or larger communities. Thought about this way, social media extends well beyond social networking sites, messaging, and email. Even reminders from voice assistants are used interpersonally. Take the example of a father who has trouble enforcing time limits with his young kids. When he simply tells them that they've got five more minutes of playtime, the end of that time seems subjective. He may announce that time's up at five minutes or it may be fifteen. Regardless, his daughters resist and wiggle for more time. Lately he's been calling in reinforcement, saying to his home device, "OK Google, let's set a timer for five more minutes of playtime." This makes it official. When the timer goes off, he is not the bad guy.

A related exchange played out between two brothers who had grown apart. The younger of the two, Paul, recently sent his brother, Roger, a voice message, through the Alexa app, in which he sang a song they laughed about as kids. This endearing exchange took an amusing turn when the smart speaker in Roger's kitchen, which was playing Paul's message, apparently took Paul's singing as a cue to play the original version of the song. Roger was suddenly transported decades back, to a time when he and his brother often laughed together about this silly song.

Rather than regarding technology as an external force or temptation that we have to struggle against, I propose thinking about the alliances that we form with technology. This alliance begins when we acquire or access something, perhaps a new device, service, or data, and evolves as the technology challenges us and we challenge it. We bring the technology into social situations it wasn't designed for. We draw on it to negotiate the limitations that we see in ourselves. In exploring new applications for it, we find new perspectives on ourselves and our social worlds. As discussed in chapter 6, "Picturing Ourselves," the technology that we bond with doesn't have to take the form of a physical device. That kind of bond is less and less likely as our data moves to the cloud. And ultimately, this bond is internalized. It becomes part of us and our sociality. We are changed by this alliance.

In suggesting that we ally with technology, I am adapting the concept of the therapeutic alliance—the collaborative bond between patient and therapist that has been identified as a critical element of treatment.[15] This alliance develops in part from mutual challenging: the therapist questions aspects of a patient's life story that may limit her expectations for the future; the patient critiques the therapist's interpretations and the process. I suspect that the value we get from technology similarly depends on how we challenge it and let it challenge us.

Many of the most successful alliances with technology that I've observed come about as people "misuse" it. These subtle departures from the prescribed uses of a technology are part of how we own a device. These hacks or off-label uses take many forms. They sometimes involve sharing a device that was designed for one's personal use, as Chandra did with the Mood Phone. Or it can be the opposite: using a social tool like Tinder for personal validation, as explored in chapter 5. The commonality across the stories in this book is individuals' adaptation of technology to meet their own objectives. Their significance is personal and interpersonal, in contrast to the politically and economically significant examples described in prior scholarship on appropriation.[16] Most of the examples in this book are simple, inconspicuous actions that support psychological well-being and intimacy.

How to Read This Book

You can approach this book as if it were a gallery. There is an order, but it is also fine to wander.

Each chapter is a room, a collection of stories organized around psychological themes. These themes pertain to intimacy and conflict negotiation, evaluating and representing oneself, remembrance and loss, habit change, identity fluidity, and managing emotional challenges and physical illness.

Each story is a piece, depicting an individual's creative use of technology. Some find novel ways to signal affection and frustration, as you'll see in the first chapter. In the second chapter, you'll meet people who tweak their digital behavior to enable difficult conversations. Some individuals track themselves and their interactions. You'll find them in the third chapter, and in the fourth, people will struggle to remember and forget. If you've ever wondered what other people are thinking as they navigate dating apps, you'll read about that in the fifth chapter, and in the sixth, you'll see how technology mediates identity. The sharing economy has altered our interactions with strangers in a number of ways. The seventh chapter will reveal some of them. And in the eighth, you'll find examples of how some people take technology into their own hands in times of despair and illness.

I wrote and framed these stories, but these individuals can be thought of as artists. Through subtle manipulations of technology, these individuals allow shifting perspectives. They open up options for how they see themselves and how they are seen by others.

Ideally, some of these stories will resonate with you. It may be that you share their goals or use the same technologies. Perhaps you too are in a long-distance relationship. Maybe you also aspire to snack less. You too may be bemused by alerts from your smart scale or glucose monitor. Or you might identify with something else. You may also dislike being nudged to do something that you already enjoy or find that talking things out isn't always helpful. Maybe you too are trying to change your health by changing your home environment.

More than likely you are already using some technologies differently than they were intended. You can probably imagine reading your own story in this book. Maybe one of the stories will help you reflect on your current use of technology and inspire you to further your own experimentation.

For readers who work in technology development, perhaps these examples will inspire designs and design processes that address unexpected uses of technology. My examination of the ways people adapt technology draws on the writing of Edward Tenner, who described unforeseen consequences of technical advances in his seminal books, *Why Things Bite Back* and *Our Own Devices*. Many of these unforeseen consequences, or "revenge effects," are negative, such as the rise of treatment resistant bacteria from antibiotics. The stories in this book resemble Tenner's description of "reverse revenge effects," unexpected benefits of technology that arise when it is adopted for a reason other than its intended use. Whether positive or negative, unexpected effects can inspire innovation when a collaborative relationship forms between developers and users. The accounts of adaptation in this book may spark ideas about how technology can support the nuanced personal objectives of end users. In addition, I hope that these stories will prompt design thinking about how people construct the significant relational and emotional themes in their lives, and how they see themselves over time. Too often tech designs focus on discrete tasks without considering individuals' broader objectives.[17]

You may wonder why this book is a set of stories. When I look back over the studies that I have done and those that my colleagues shared with me over the last fifteen years, the stories are what stick with me. The frameworks and conclusions derived from studies may influence specific academic discussions or product directions, but the individual instances have, in my opinion, a more enduring and broader value. We learn from other people just by observing them. We learn from witnessing their struggles as well as their successes.

These stories emerged from interviews that I conducted during formal studies, conversations with colleagues, friends, and acquaintances, and secondary research.[18] In all cases, my goal was to capture the individual's perspective, insights, efforts to bring about change. In most cases, the individuals were trying to change themselves, and in other cases they were trying to change a family member or policy.

In writing and drawing together these examples, I took a cue from the case study approach that was part of my training as a therapist. The case study provides a detailed portrayal of an individual or relationship—sometimes capturing how individuals view themselves. Some case studies illustrate already understood phenomena, while others describe new observations that can be explored in future research. The examples in this book are shared not to explain what is already known or provide recipes for solving particular life challenges. Instead, they are intended to invite reflection and experimentation.

Whether we are thinking about the tools that are already familiar to us or those on the horizon, the value that emerges will come not just from the technologies themselves but also through creative tailoring of them to our own needs. In the stories that follow, individuals do just that: adapt technologies to address concerns that are important to them. These are not generic hacks but personally meaningful adaptations motivated by self-reflection. Their stories invite us to reflect on our lives and our devices—that is, to engage them as supportive allies in our quest for connectedness and well-being.

1 The Meaning of Light

When the light in her living room turned on, Liza knew that her daughter, who lived miles away, had returned home safely from work. And in that moment, she felt that they were connected. Liza's living room lights turned on because they were linked to motion sensors in her daughter's home. This was part of a system that my Intel colleagues and I tested in 2003 with older adults and their remote family members.[1] We thought that the light would provide practical information, such as clues about convenient times for family members to call each other and reassurance that each person was going about their normal routines. But it turned out that the light was conveying more than that. In addition to these practical cues, it gave "a warm vibe," as Liza described.

The light did not just provide information. It sparked a visceral connection. Whether through text, light, or speech, how we reach out shapes what we convey. It has long been understood that light has social properties, that it creates spaces for interaction.[2] The light provided by fire, combined with warmth, is such a powerful force for drawing people together that the hearth and bonfire have become equated with communality.[3] Electric lighting has shaped societies and norms of urban life, as evidenced by the extreme consequences of blackouts—which, as scholar David Nye chronicles, have ranged from heightened communality to violent riots: "In a blackout, the electrified city is faced with a darkened twin of itself."[4] While it is likely that technology such as electric lighting reflects existing cultural dynamics, I maintain that we can also shape our relationships through the intentional use of light as a form of communication.

Connected devices, which now include thermostats, light bulbs, and speakers, can be linked to routers and controlled remotely by apps. They promise to save energy, effort, time, and money, just as the sensor-controlled lights developed for the project described above promised to optimize communication and convey practical information. But we are not just information-processing machines. We may initially procure connected devices for efficiency. But we use them relationally—to send and receive social signals. And sometimes these signals are more about visceral connections than coded messages.

In 2016, in a project called WeLight, my colleagues and I at the University of Washington returned to this challenge of how to support communication with light—this time with connected devices instead of the fragile sensor networks that we struggled with in 2003.[5] We wanted to enable people to send light messages across any distance. It was clear that we needed to create a tool for this since even the latest generation of connected lights blocked people from controlling lights other than those in their own homes through a strict matching of accounts to households. For example, the app for my smart lights lets me control the lights in my apartment, but not those in my mother's house. So we developed WeLight, an app that allows people to change each other's lights through simple text messaging. Users first register their home lights and phone number to WeLight. They can then write text messages with phrases like "4th of July" or "pink," and the recipient's lights will change to a color pulled from either a color library or Google image search. If I wanted to say hello to my mom some morning, I might send her a text message, using WeLight, that reads "amazing sunrise here," and her lights would turn shades of pink or other colors found in pictures of sunrises. Texting a picture of the sunrise would have the same effect. Words and images are transformed into colors that shine throughout a recipient's home. Instead of supporting a functional task, lights become ambient messages. And rather than reading a text message, the receiver is immersed in light.

Leading up to WeLight, I interviewed individuals and households about "smart" lighting. I wanted to understand their environments, their relationships, and the problems that they were hoping the technology would help them solve—in other words, the contexts in which they would use the app. I then asked people to play with incorporating smart lights into their communication and interviewed them a month later. As we'll see in their stories, the lights were used for emotionally rich signaling. And in addition to signaling affection, lights became a way for people to nudge each other, convey practical information such as location, and work through conflict, as we'll see with the first couple.

Using Light to Throw Shade

Sasha and Nick, a couple, have strikingly different personalities. Sasha is reserved. Her arms are often folded, and she's typically trying to avoid unnecessary interaction. Nick has boundless social energy. His gestures are big, his clothing is bold, and he is always making new friends. One of his frequent social habits, until recently, was bringing home dinner guests unannounced. Sasha isn't crazy about impromptu hosting. Even if she likes the guests, she resents the lack of opportunity to veto the plan.

When Nick announced one afternoon that he was going to show up with a new dinner guest in tow, Sasha thought about how she could change this pattern. She was still at work when she received Nick's message. She didn't want to confront him in front of the guest. So before coming home, she experimented with the lights. From her office, she turned one of the smart lights in their house deep red, knowing that he would see it immediately when he got home. The next week, when the same guest reappeared, she turned three of the lights red. Nick finally got the hint. The lights cued him to her feelings without putting him on the defensive in a way that a verbal confrontation might have. After the guest left, they sat down together with the app, exploring other color ranges, and discussed mutually agreeable ways to host guests. Through the lights, they externalized and modulated the conflict, setting the stage for constructive dialogue.

Smart light manufacturers do not suggest that you buy their light bulbs for the purpose of working through conflict. But light—whether from a fire, the sun, or a bulb—has powerful metaphorical value that can enhance communication.[6] We can use colors and other qualities of light to convey urgency or something else about how we are feeling. For example, darkness and brightness are often invoked to understand depression and recovery.[7] We can also use light as a signal of how we would like to feel, what Stanford University psychologist Jeanne Tsai calls our "ideal affect," or how we would like things to play out with others.[8] Sasha used lights not just as a complaint but as an invitation to change a dynamic. The message, as Sasha and Nick started exploring other colors within the app, was something like, "Just as we can change the lights, we can change this friction between us."

Many self-help books about relationships would tell Sasha to sit Nick down and open up a dialogue about her feelings. "Communication is key," we are told over and over. And that is true. But direct, verbal confrontation is not always the best way to initiate communication, especially when one of the partners is angry. Often, anger comes out as an accusation that the other person did something wrong or is inherently flawed. Along with the obvious damage to relationships, anger can jeopardize cardiovascular health over time.[9] Instead of angry confrontation, negotiation experts advise "getting on the same side of the problem," sometimes by literally sitting alongside the other person so that the two opposing parties can confront a problem together.[10] The idea is to turn an opponent into an ally and collaboratively solve the problem.

In the example above, Sasha did just that, using the lights to externalize the conflict. She created a physical embodiment of the conflict to avoid suggesting that the problem was Nick himself. And this symbol of the conflict allowed them to recognize its impermanence. The dynamic qualities of light suggested that their conflict was similarly changeable. They were both interested in seeing the red change to green or blue. Red light had an obvious meaning to Sasha, akin to a stoplight, and that meaning was apparent to Nick. Their color signaling of course differs from the sexual invitations of red-light districts. In this example and others, the context of the relationship provides meaning.

Now you're probably wondering why Sasha didn't just say, "Stop bringing people home without asking me!" In fact, she had said those very words many times before, for this was far from the first time that Nick had invited friends over unannounced. It may have been precisely the spontaneity of these invitations that appealed to him. It is hard to change behaviors that are expressions of extroversion, openness and other personality traits. A new approach breaks the cycle—casting new light on it, so to speak.

Present in Absence

It seems like the phone would be a desirable way to keep in touch for couples who spend time apart. But that hasn't been the case for some couples I've interviewed. When one was traveling and the other was at home, phone calls failed to provide the connection that they were seeking. Rather than feeling closer during these calls, they became emotionally disconnected.

Their voices were transmitted clearly, but their circumstances were either not appreciated or not conveyed. Some described conflicting urgencies: in one couple, the traveler's desire to squeeze in a family video call before a business dinner posed yet another burden for the at-home partner, who was juggling playdates, dinner, and baths. In some cases, when the travel itself was a source of tension, the call intensified the feeling of distance. When conversations were strained, some couples, like Elana and David, found other ways of connecting over a distance.

Elana is not a gadget person. She has minimalist sensibilities and wouldn't normally spruce up her home with networked appliances, but she allowed her boyfriend, David, to install smart lights. David lived four hundred miles away from Elana—a separation required by their jobs. The distance was a major source of tension in their relationship. He often got upset about it and refused to talk by phone. She, in turn, worried about him and their relationship.

So when she came home one day to find her apartment lit up in violet and blue, she was reassured. David had access to her connected lights, and this was a message from him. She felt loved despite the tensions.

To enable this remote control of lights, David had linked his phone to Elana's lights during a previous visit when they'd set up the lights. They each wanted to be able to control the lights in each place while there or away. When David wanted to adjust Elana's lights from afar, he signed into her account and selected colors for specific lights. He knew approximately when she'd be arriving home and could change the lights from any location. This involved some challenges, particularly disambiguating the accounts after adjusting lights in one another's homes. Sometimes Elana couldn't reestablish control over her lights from within the app and had to either text David to turn them off or physically unplug the lights when she wanted to sleep. Nonetheless, the light exchanges became an important part of their long-distance relationship.

How could turning on lights come to mean so much? In part, it is because we have a long history with light. We associate it with the warmth of fire. As Lisa Heschong writes in *Thermal Delight in Architecture*, "Are the colors reds and browns? Then maybe it will be warm like a room lit by the red-gold light of a fire."[11] We associate fire with interpersonal as well as physical warmth. Light has always brought people together.

And we have long used light as a signal. Signaling theory tells us that organisms have evolved so that everything from the colors of a butterfly's wings to a human's casual gestures can convey essential information. Take the male goldfinch's plumage, which transforms to a vibrant yellow in mating season. The bright colors, which reflect carotenoids and immune functioning, are an "honest signal" of health to potential mates.[12] From Microsoft research comes a contemporary version: makeup that changes color to signal changes in air quality. Eyelids that turn from silver to black, or lips that go from pink to white, may warn passersby that this is not the place to take a run, or may provoke someone to confront a smoker. Collectively, these skin signals could function as a protest against pollution and the depletion of the environment.[13]

Signaling theory is often concerned with deceptive signals: animal markings such as "eye spots" that confuse predators, or, for a human example, the comb-over. But Elana and David were using lights, which have nothing intrinsic to do with love, to convey a truth. And these messages, unlike a text, don't divert their attention to a phone; they are immersive.

The medium need not be limited to light. Many familiar objects have been explored as channels for intimate communication—clothing that "hugs" over a distance, teacups that glow when picked up at the same time, or a mattress heated in the spot of a faraway significant other.[14] Recent work on "ghosting" took this a step further, synchronizing the lights and sounds in two homes.[15] Body-to-body connections have also been prototyped, for example, through actuators that move an individual's arms in gestures according to the mental state of a long-distance romantic partner, as measured by an EEG.[16] These examples involve passive sharing of sensed behaviors and biometrics, but it may be more powerful to actively create experiences as a form of gift giving. We will doubtlessly find more immediate and compelling ways to share music and scenic views over a distance. And I suspect connected devices will also support more tangible forms of expression, perhaps allowing us to make a friend a cup of tea or even dinner from far away.

Ambient Nudges

In my research on how people communicate via connected lighting, I've found that the uses extend beyond the signaling of emotion described in the first two stories. In the example below, lights are used as an intervention to nudge someone else into behavior change.

Karthik, a twenty-four-year-old engineer, lives with his parents and twenty-two-year-old sister outside London. He appreciates his parents and tries to cultivate a warm family vibe. He typically leaves his bedroom door open and spends time with his parents in the common areas. His sister, in contrast, often locks herself in her room with her headphones on. To draw her out of her room, Karthik experimented with smart lights.

First, he put the lights next to their childhood pictures on the wall in the family room. From their rooms, he and his sister could signal their availability to their parents by changing the color of the light bulbs. His sister always left her light red, conveying unavailability.

Next, he put one of the smart lights in the desk lamp in his sister's room. She could control the light from her phone, but he was able to override her selection. At dinnertime, and other moments that the family gathered, he and his parents turned on her desk light from their phones to request her presence. Even those not so subtle prompts were ignored.

He then escalated the intervention, installing the smart light in place of her primary overhead bedroom light. The signal he chose was to blink the light, making his sister's room intolerable for her.

Immersed in these new light invitations, she finally joined the family. After a week, she demanded that her brother swap out the light, which he did. It might surprise some readers that this intervention, as intrusive and annoying as it was, lasted even as long as a week. But families are idiosyncratic, and my impression is that Karthik's sister was unusually adept at blocking out people and unwanted stimuli. In any case, the short-lived experiment interrupted her reclusive behavior and sparked some family discussion about their differing desires for interaction.

Karthik's sister had been unwilling to exert any energy to make her parents feel appreciated. She refused to use the simple signal that Karthik initially set up for her to convey availability to her parents. She ignored the slightly more intrusive desk lamp signals that invited her to dinner. But she did respond to a more immersive nudge to dinner when the signal became more than that and overtook the overhead lamp's primary function of shedding light in her room.

This progression was not as linear as I've portrayed it. Over the course of a month, Karthik experimented with seven different ways of using the lights to improve family dynamics and found several that worked, although some were effective only for short periods of time. Signaling can go awry even when we are using an acknowledged signaling system, such as language, if there is not a shared understanding of meaning and desired actions. This agreement may be even more critical when appropriating technology, stretching it beyond its original purpose, in order to signal meaning.

Karthik's experiments underline the importance of shared goals when trying to instigate change. A signal should reference these shared or complementary goals, just as verbal negotiations should.[17] Karthik's sister didn't value communality in the way that her parents and brother did. She was most concerned with personal peace and privacy. She complied with the light prompt when the flashing interfered with those goals, but as soon as she figured out how to reestablish an insular bedroom environment, she did just that.

In addition to the message about dinner, the light in this case conveyed negative feelings, especially frustration and hostility. Just as the intent of human touch can be inferred in the absence of other cues, the feeling behind ambient messages comes across along with the message itself.[18]

One of the principal flaws in Karthik's intervention is that it lacked reciprocity. Even if his sister had wanted to participate, she wasn't on a level playing field, for in this household, as in so many others, there's one tech-savvy person on whom the family relies. It's good to have an in-house technical expert, but this exclusive control can close down opportunities for others to control their environment in basic ways. Feeling in control of our environments affects how we feel about those places, our emotional states, and our ability to get things done.[19] That perceived control cultivates self-efficacy: the confidence that one can undertake the actions necessary for a particular goal along with the motivation to take those actions.[20] The opposite—an absence of control or the perception that one is personally not capable of bringing about change—leads to learned helplessness.[21] If convinced their efforts will be ineffective, people generally stop trying to escape aversive situations. Learned helplessness can spiral into passivity and depression. In the case of Karthik's sister, a realization that she was not able to control her own lighting could lead to further disengagement from home life, exasperating the very problem that her brother and parents were trying to solve.

This example also makes it clear that signals are rarely just a way to transfer information. We can easily cross the line from immersive, emotionally rich signaling to aggressively controlling someone else's environment. Devices that serve nonsymbolic functions, such as a blinking overhead light, can have a profound and sometimes negative effect on others. The line is fuzzy. We should cross it carefully.

Taking Control of One's Signals

Ambient signaling can also restore control to people who need it. One man told me about the challenges of communicating with his autistic son, who cannot speak. His son is extremely sensitive to environmental cues and frequently assigns his own meaning to them. One of these cues involved light. A malfunctioning red light on a building outside caught his son's attention. He could see it from his bedroom window and read it as a sign that he should stay inside. Every morning, he screamed in protest as his parents woke him and prepared him for school. Every morning, they would calm him down, assure him that all was right, and eventually get him to school. They couldn't do anything about the red light but wanted to give their son a greater sense of control over how he interpreted the environment.

His father thought that connected lights could help. His son could use them as a distress signal; if anything was wrong, he could alter the lights in his parents' room through the app. And his father would set the lights in the son's room to reinforce waking and sleeping routines. Part of the beauty of this idea is that it is a relational cue: his son would likely associate the light message with his father. In contrast to a buzzing alarm clock, the light might engender greater trust and feelings of being nurtured.

I don't know if things went as the father imagined since he was not available for a follow-up interview, but his plan offers ideas for how we can redistribute control in a household.

Weighty Effects of Lightweight Symbols

In the examples above, light was used to work through conflict, signal emotional presence across a long distance, nudge a sister out of her bedroom, and calm a fragile child. In the stories that follow, signals that are generally considered to be lightweight forms of communication facilitated nuanced, weighty interactions. One group of friends found a particularly creative way of lightening communication that had been deemed inappropriately dark.

Fight Club, without the Clubs

Regina and her friends, all enrolled in an MFA program, openly expressed rage about their department, dating, and other aspects of their lives. They didn't hold back; often, they playfully threw furniture around their studio, turned tables upside down, and literally screamed their frustration out the windows. Not surprisingly, this catharsis drew concern from their school.

Over time, they found a replacement activity: creating a new symbolic language to depict this physical aggression. Combining lines and symbols available on their mobile phones, they illustrated combat, upside-down tables, and screams. They were, for all intents and purposes, creating their own system of emoji. Rather than the typical use of emoji as emotional shorthand, though, Regina and her friends used their homegrown emoji to express emotional nuances.

Their anger was social—something that they experienced and, in their own way, celebrated together. One of her friends manifested this by combining symbols to represent a boxing match. She sent it to her friends with the text, "Wanna fight?"

Many fights ensued in their ongoing competition to see who could most quickly create the strongest depiction of a particular feeling, be it frustration, indignation, disappointment, or hopeful anticipation. Their signals were not intended to communicate efficiently. For that, they could have drawn from the extensive libraries of emoji that have become integral to communication. Their creative assemblages bear some resemblance to ASCII artists, who use keyboard symbols to draw images or even modify the appearance of the Twitter text box.[22] Emoji are often used to reassure others and provide a kind of scripted emotional labor.[23] But Regina and her friends put emoji to a different emotional and relational use, pronouncing their rage rather than trying to smooth things over.

The painstaking effort that Regina and her friends put into the emoji was itself a signal, conveying both the intensity of the feelings they represent and the successful sublimation of those feelings into acts of creativity.

Iron Fist in a Velvet GIF

Assertive communication is complicated for anyone with any social sensitivity, and especially hard for women—not because they can't be assertive, but because when they are, there's frequently a backlash.[24] That's why some experts advise women to combine warmth and dominance—"an iron fist in a velvet glove," as the saying goes.[25] Or as Sheryl Sandberg puts it, less dramatically, women should "ask for promotions (with smiles on our faces, of course)."[26] The "of course" signals just how ingrained these biases are.

I've been impressed by the ways that some women use social media to convey sensitivity in assertive communication.

Fiona, a graduate student in sociology, described using GIFs (animated images, often of relatively poor quality) to diffuse tension with the male colleagues who were helping with the software development for her project.

Before a huge deadline, one of the developers sent her a cute image of a cat saying, "Roger, we are go for takeoff." With this LOLcat, he was letting her know that he thought they were all done and ready to launch the interactive application.

At the eleventh hour, though, Fiona realized that she needed an important change in the code. That meant asking the developer to rework things significantly. She prefaced an email with another LOLcat, "One does not simply apologize with an LOLcat," referencing a line from *The Lord of the Rings*: "One does not simply walk into Mordor." The developer's first message had signaled his familiarity with LOLcats, and her response announced that they had that and other references in common. In her words, "We were having this really charged emotional back and forth at this crucial time, and the LOLcat pics just helped us to communicate being on the same page, using humor, defusing big emotions, communicating gratitude, and speaking each other's language. There was reciprocity there."

Exchanges such as Fiona's with her colleague are so commonplace that it would be easy to look past them. But if we pause to look closely, we see that these lightweight symbols carry significant, nuanced social meaning. Every signal conveys more than its explicit content. If I tap a message to you in Morse code, I am also signaling that I know Morse code and think that you do, or should, too. If, as we saw earlier, an older brother like Karthik blinks the lights in his recalcitrant sister's room to get her to come spend time with their parents, he signals that he cares about her and their parents, and likely also that he is so frustrated with her that he is controlling her environment against her will. And, every signal implies a shared language—even if it's a nonverbal one. In Fiona's case, reminding the developer of their commonalities outside the project was a crucial part of the message. She integrated these shared cultural references with the literacy required for a meme to go over well.[27] Like others who skillfully draw on the anthropomorphized cats, Fiona comes across as accommodating and funny.

Fiona's instruction to her colleague could have been blunt and devoid of social finesse. Instead, her reference to their shared cultural affinities along with her consideration for his expectations made it clear that they were working toward a common goal.

Finding Meaning in Light

Light can mean different things. As explored here, lights can convey substantive feeling and meaning just as emoji and GIFs do. These symbols are lightweight but not trivial means of communicating. In large part, we use them for relating on an emotional level, for what anthropologist Bronislaw Malinowski called "phatic" communication.[28] Think of a wink, a wave, a pat on the back, or the modern equivalent of a Facebook "like." These expressions convey emotion and help support relationships. The same modality can be used for either phatic or instrumental communication, where there is a meaning to be decoded. Take the example of missed calls (common in India and Africa in the mid-2000s), where to avoid fees, callers let the phone ring a set number of times before hanging up. Communications researcher Jonathan Donner, who studied this practice in Rwanda, found that missed calls were sometimes used relationally—for instance, as a way of saying "I'm thinking of you"—and sometimes instrumentally—to signal something like "I'm done and need a ride home."[29] The meaning is drawn from previous conversations and an understanding of the other person's situation; the relationship provides the context for understanding the message.

In the stories above, connected lights conveyed feelings such as reassurance and anger as well as information like location and dinnertime. Often, both feeling and information come through. Karthik's sister doubtlessly sensed his impatience in the flickering lights that beckoned her to dinner. And a boy awoken by the lights programmed by his father may feel that caring as he realizes that it is time for breakfast. Lightness is also at play in the LOLcat GIFs, which diffused tension among collaborators, and homemade emoji, through which a group of friends channeled anger into shared artistic expression.

Many of the technologies discussed in this chapter are inherently communicative. Emoji were meant to be sent, and light has always enabled interaction. These and other technologies are baked into our relationships, conversations, and power structures. But as the examples above show, we can enhance and personalize the communicative power of technologies. We can tailor them to our relationships along with our goals for those relationships.

The type of communication portrayed in this chapter will become easier and more nuanced as our devices become more aware of our emotional and physiological states. There will surely be features that automatically adjust music, light, and other ambient conditions to our mood. But I contend that such adjustments will have value only if people are in the loop. Perhaps friends and family members will grant each other access to such personal data and ambient controls in order to express care by adjusting one another's home atmospheres.

2 Conversational Catalysts

Some conversations get stuck, and some are hard to start. It can be difficult to inquire without prying, see from the perspective of someone we are arguing with, or find a way to soothe a child who is upset. In the stories below, people experiment with technology to address such challenges. Some of these conversations are among people who know each other well, and others occur in new relationships and even on first dates. These individuals make use of digital traces, games, crowdsourcing, and self-help tools, applying them in slightly unconventional ways. But in all cases, the technology is just a point of departure. These stories raise the question of how, in many other situations, we can use technology as a bridge rather than as a surrogate for conversation.

The Big Mouth Scale

Miguel is an exuberant single man in his late forties who often hosts parties and guests. One day, when he was hosting an out-of-town female friend for a few months, he was surprised to receive an email from his scale congratulating him for losing fifteen pounds. This initially surprised him because he knew that he had not lost any weight. Then he realized, a la "Goldilocks and the Three Bears," that someone else had stepped on the scale in his bathroom. This person must have been close to his own weight, which ruled out his girlfriend or his female guest. He realized that his guest must have had a male visitor.

He sent her a teasing email, congratulating her on her new romance. She teased back that Miguel was confused by the scale and didn't understand modern technology. They both knew of course that this was all a joke, but the playful banter opened up a more genuine conversation about the new mystery man and her love life.

Miguel frequently hosted guests for extended stays, not for rent contributions, but to create a sense of family and home life. The way that he used the data from the scale is aligned with this intent: it sparked the familial dialogue and closeness that he was seeking.

Miguel's internet scale is intended for its owner. Its communication and social network capabilities are meant to help the scale owner stay on track with personal health goals. Miguel expanded on this intended use, drawing on the errant communication to enhance a relationship. The misdirected congratulatory note to Miguel clued him into his houseguest's romantic liaisons and allowed him to deepen their conversation.

The data from our devices connect us to others in ways that we aren't always expecting. There was a breach of privacy through the scale, one that Miguel playfully leveraged for conversation. The breach of the visitor's personal data is not shocking given that it was Miguel's scale, in his bathroom. But it may strike a nerve. For many, stepping on a scale is a personal act. They are not intending to broadcast that particular action, which itself can signal weight concerns, let alone the readout of their actual weight. Traditionally, scales have captured just a snapshot. Until recently they haven't been computational or connected; it probably doesn't occur to most visitors that their data will become part of a trend that is displayed to a device owner, and perhaps that person's social network. Connected scales can cause embarrassment, deliberately or accidentally. They can even expose an affair that one was trying to conceal.

Nevertheless, when data escapes, there may also be some opportunity for productive conversations. As explored in several of these examples, data from personal health tools may provide a window to express interest or concern.

The Accidental Sleuth

Juan is a fidgety engineer with dark glasses. His constant movement and Mediterranean diet keeps him thin, and he has never had any major illness. He isn't overly attentive to health, but he started to notice weight entries on his smart scale that were taken in the middle of the day, when he and his wife were at work and the kids were at school. There was only one person who would have been home at those times. He quickly deduced that his housekeeper, Lucinda, was weighing herself before and after eating lunch. At first, he found it mildly amusing to calculate how many grams she consumed for lunch each day, but he became concerned about a visible trend of weight loss. He worried that she might be sick. His fears were confirmed. She was hospitalized with severe stomach problems several weeks later. Fortunately, the condition was treatable, and she recovered fully.

It's possible to read this as a cautionary tale about stepping on scales or other connected devices. Scales are hard to resist. They immediately generate a number that we are trained to read as a proxy for health, attractiveness, and personal responsibility. As shown above, it can become complicated. Our number can become part of someone else's story. When stepping on a scale at a hotel or a friend's home, we may not be the only ones seeing, talking, or tweeting about our weight.

But the story also raises questions about how data from connected devices becomes embroiled in how we care for each other and how we communicate about accidental breaches of privacy. Juan was relieved that Lucinda was okay, but he was consumed with regret for not raising concerns that could have prompted earlier treatment. He was embarrassed that he initially derived some entertainment out of watching her scale entries and had been too ashamed of his spying to raise his concerns about her health. How could he tell her that he'd been monitoring her weight for weeks?

In retrospect, Juan wished, as I think many people would, that he had raised his concerns immediately. Yet to act in this way would have involved acknowledging the type of transgressions one typically ignores—in this case, that Lucinda had used Juan's personal device and that Juan was reading the scale's results. He worried about her reaction to being watched—not by government or a big company, but by him, her client. Even if she appreciated Juan's concern, the disclosure might also have made things awkward for her. She would know that Juan had in a sense watched her without her awareness. Perhaps there will eventually be a common preface for such discussions—a shorthand for "I know I shouldn't know this, but since I do, I feel obliged to tell you that..." when we need to relay information to which we never should have had access.

Words with Socially Anxious Friends

While Juan concealed his awareness of his housekeeper's data to protect her feelings of privacy, Ciel circulated information to bring a withdrawn family member back into the fold.

Ciel is in some ways the matriarch of her family. She watches out for everyone and provides the glue in her extended family. She financially supports her son, mother, grandmother, sister, and nephew, and she also helps care for her extended family in nonmonetary ways.

Ciel never knew her father. She escaped an abusive husband shortly after giving birth and brought up her son as a single mom. She runs a successful small business. Ambitious and extremely hardworking, she thinks about her life goals primarily in professional terms. But even at work her nurturing manifests. She keeps an eye out for her employees and cultivates a strong culture of belongingness. Without feeling burdened by it, she interweaves caring for her family with work demands. On the way home from the office, before she resumes working from home, she often makes detours to help her sister, pick up her nephew, or drop off something for someone. She doesn't seem to resent this time that she gives to others or even recognize it. Her roles as mother and CEO rarely run up against each other explicitly, but after long days she sometimes collapses in exhaustion.

Ciel notices when someone in her extended family is struggling and is often the first to step in. In recent years she became concerned about her cousin's husband, Derek. Derek was notably absent from family gatherings and was withdrawing more as time went on. He was plagued with social anxiety; the longer he avoided these events, the more he dreaded attending a future one. His wife's apologies for him worsened things, raising concerns about why she was covering for him. Ciel liked Derek and felt bad that he was becoming alienated. She empathized with his work as an emergency medical technician, knowing firsthand the dangers of living in a chronic state of high alert. She imagined that he needed ways to relax as he waited for emergency calls, responded in full force to accidents, and then tried to forget the horrors that he had seen.

At one point, Ciel and Derek exchanged invitations to Words with Friends, the crossword puzzle app. Ciel cannot remember if this was Derek's idea or his wife's, but their game playing had an energy that took her by surprise. The man who was monosyllabic at family events was hyperfluent here. He

matched her words in complexity, attuning to her like a skilled dance partner. Sometimes there were even messages for her within the words of the
game. At one point they had thirteen games going on in parallel. If family cel-
ebrations were at one end of the spectrum, online word games were at the
other. His hunger for connection was obvious. The asynchronous nature of
Words with Friends may account for Derek's ease of interacting with Ciel
within the game. As game designer and media scholar Ian Bogost explains:

> None of the social anxiety of long turns exists in a distributed play session. And
> besides, each player has his or her own private screen for play, thus making it pos-
> sible to hide experimental moves in a way that wouldn't be possible on a coffee
> table ... that slow, deliberate exploration of what a game can be, what it can do,
> and how it can be shaped in the hands of its players and designers over a very
> long time—that's a virtue, and an unsung one.[1]

Derek's wife appreciated the Words with Friends connection between
Derek and Ciel. She wanted Ciel to see Derek's strengths and talk about him
in positive terms with the rest of the family. Ciel did so, trying to repair
his reputation. She also suggested that others, particularly her sister, com-
pete with Derek at Words with Friends. Derek would always win the games
with Ciel. They both had an extensive vocabulary, but he had mastered
short words too. Ciel played her best, yet wasn't disappointed about losing.
Winning wasn't her objective. Her intensely competitive sister, by contrast,
set out to win in her games with Derek and occasionally did. News about
their matches spread throughout the family, and Derek became known for
his verbal prowess.

When I last spoke with Ciel, she had backed off. At some point the intensity and intimacy of the games maxed her out. She realized that she just
couldn't spend that much time playing Words with Friends. The invitations
seemed relentless, and she wasn't sure how to dial it down. Her sister and
Derek were still playing, but she suspected that too would subside at some
point. Their use of the game provided a lifeline to Derek. It changed conversations from "What's wrong with Derek?" to "Wow, who knew that Derek
had a boundless vocabulary?" It suggested that people could connect with
Derek if they thought beyond birthday parties and other family events. Ciel
and her sister see the game as a temporary strategy—one that will be replaced
by others as Derek reconnects with his family.

When Ciel looks back at this, she sees the power of virtual connections
that simply have no analogue in the physical world. We rarely dispense

with all conversational norms and jump into intense word games when we see our cousins. But she thinks that we should be open to that kind of connection and appreciate it for what it is. Ciel brought her cousin back from the fringes by pivoting technology, Words with Friends, in a subtle but important way. Many people find that their connection with a friend intensifies while playing similar games. For most, this connection is a by-product of game play. For Ciel, it was the draw. She cares about words, but she's not looking to amuse herself or kill time. She turned the by-product into the motive.

Here, we see games at their best. Assembling letters into words, in playful competition with a caring family member, brings someone from darkness into light. In Bogost's words, "A game, it turns out, is a lens onto the sublime in the ordinary."[2]

The Bond of Gossip

Maia and Frank have been friends for a long time. They went to the same high school, parted ways for different colleges, and now live in different cities. Over this long geographical separation, they've sustained their connection through the Facebook page of someone they barely know.

In high school, they had a mutual acquaintance, Jenna, of near-celebrity status. She had thousands of friends and frequently posted outrageous updates and pictures. She was almost what media scholar Teresa Senft termed a "microcelebrity."[3] She projected emotional volatility and sexuality to cultivate an audience. She had, like the microcelebrities studied by Professor Alice Marwick, edited herself into an engaging consumer product.[4]

Even now, roughly five years since their high school graduation, there is still major drama and warfare on Jenna's Facebook page almost every day. Maia and Frank have continued their diligent tracking of Jenna's Facebook posts and videos because she frequently deletes them. "Video now!" is their signal for a video that is too good to last.

It's a soap opera they've watched for almost a decade. They have tracked Jenna's breakups, partying, trips to rehab, sobriety, pregnancies, moves, and her parent's divorce. Occasionally, there have been mysteries to solve, such as when they discovered that she had multiple Facebook profiles.

Beyond entertainment, Jenna's Facebook page has served as a way to smooth out the rough patches in Maia and Frank's relationship—a conversation starter that reminds them of their own history as friends. When one has been angry at the other, screenshots of Jenna's status updates have typically cleared the air. Maia remembers only one time that it didn't do the trick. She knew then that Frank was really upset and a genuine apology was required. She happily gave it.

This is, of course, gossip, but it is not particularly pernicious. It is similar to bonding over an always-in-trouble celebrity, but here the conversation starter is rooted in their shared past. And they aren't just rehashing their high school days. Jenna is always generating new content for discussion.

Gossip is understood as a crucial bonding mechanism and common form of communication.[5] It is a way of exchanging socially relevant information about problematic peers, or those with fame who function as either positive or negative role models. But someone like Jenna garners significant attention without the protective resources of major celebrities, such as a public relations team to ward off criticism.[6] She builds interest among strangers in part by exposing her vulnerabilities and sharing intimate details of her life. Such performances of authenticity and intimacy can draw criticism from strangers and may alienate others.[7] This example raises questions about the social use of someone's provocative posts. Maia and Frank routinely bonded over Jenna's Facebook activity, but that was never where their conversation stayed.

Sharing a Mood Map

Some rifts are harder to mend. Tobias, an anxious family man with a demanding job, had struggled to communicate with his wife for years, and the distance between them seemed to be growing. They had different lives and swapped rather than shared what they had previously created together: their children and home. At precisely timed schedules, they exchanged responsibilities and locations. In our first conversation, Tobias focused on his own feelings of resentment.

I met Tobias because he volunteered to try the Mood Phone, an app that my colleagues and I designed to help with stress.[8] I talked with him every week throughout the monthlong study. One element of the app, a Mood Map, appeared on the phone every hour. It asked him to indicate how he was feeling at that moment (that is, how energetic and how positive he felt) by dragging a dot to a spot on a two-by-two matrix. Based on his current mood, the app launched other scales and therapeutic exercises, including breathing visualizations and cognitive reappraisal prompts.

Tobias noticed a pattern in the data: his mood dropped every night when he came home from work and typically remained low the whole evening. The transition home was rough; as soon as he walked in the door, his wife rushed out to the gym, the dog jumped on him, and the kids demanded his attention and dinner. The daily pattern contributed to his resentment.

Over time, his reflection shifted from resentment to curiosity. He became interested in how his wife was feeling. Tobias wanted a way to ease into more empathic conversations but worried that directly asking her "How are you feeling?" would come across as a confrontation. He imagined her firing back, "Why are you asking?"

If his wife also had the mood-tracking app, and if their phones could exchange information about their moods, he thought, maybe they could use their shared data to open a conversation more naturally. Their phones weren't able to sync in this way, but his curiosity alone helped him open some of these conversations.

Today, many people wear or carry at least one device that tracks an indication of mood, even if that is not its primary purpose. It may be a mood app on your phone, like the one described here, or sentiment analysis of your posts in a social networking application. Or perhaps you wear a wristband that monitors heart rate variability and electrodermal response. MIT Media Lab professor Rosalind Picard, who introduced affective computing and founded start-ups Affectiva and Empatica, has shown these to be measurable signals of stress. These metrics signal arousal: the degree to which you are "amped up." They do not give as many clues to valence: the negativity or positivity of your mood.[9] Relying on arousal signals alone, it can be hard to distinguish between excitement and anger, for example, or make sense of mixed feelings. By analyzing measurements in combination (e.g., of physiology, facial expression, and language), our devices might give a better estimate of our emotional states.

But to have a lot of value, our devices really need our input. Devices just display a best guess based on data that is often sparse, sensitive, and nonspecific. The endgame is not having a device that is smart enough to tell us how we are feeling. Our emotions are different from things like temperature or blood glucose level, which can be measured independently of how we experience and describe them. As explained by Lisa Feldman Barrett, professor of psychology at Northeastern University, emotions take form as we interpret events and our physiological states. The richer the repertoire of emotional concepts we have to draw on, the more precisely we can name our feelings. This articulation shapes our experience of the world: the more precisely we can label a challenge, the more effectively we can respond.[10] Feeling "bad" differs from articulating "righteous indignation," Barrett points out; the latter is more likely to propel one into action.[11] "Emotional granularity" creates more options for understanding and reacting to challenges. This ability to finely articulate emotions will likely also help us understand and relate to others.

An app isn't a conversation substitute. But it can help open the conversations that we want to have. Tobias tracked his moods, noticed a pattern, and wondered about his wife's moods. This prompted him to initiate different kinds of conversations with her and express interest in how she was feeling. Their dialogue improved. He wasn't alone in finding this kind of benefit. The participants who described the most benefit from this mobile intervention were those who brought it into their conversations—with family, spouses, and in some cases colleagues. With emotional tracking data, some may find value in sharing with a spouse or another close tie, while others may benefit from relating to peers, perhaps those with a common concern or demographic.[12] When there is an opportunity for this kind of sharing, connection will come from trying to understand the nuances in the other person's experiences.

Emotional Rough Drafts

Tobias needed a way to start a conversation and break through tension in his marriage that had built over years of silence. Matthew had the opposite predicament. He needed to stop some conversations and slow down others that could quickly overheat. At the start of our first conversation, Matthew was all business. He spoke in a dry, matter-of-fact style. It took a while for him to open up in our conversations, and he said that this was true for him in many situations. He tries to express himself in a calm, steady manner, in part to avoid expressing the intense frustration that he often feels. In the past, when he has let his annoyance surface at work, it has damaged some of his relationships. As Matthew's role shifted from contributing individually as an engineer to managing a team, it became even more important that he communicate with care.

He knew that email was a danger zone. He needed to slow down his communication, to catch the anger pounded out on his keyboard before it reached other people's in-boxes. So he set up a delay between the time that he hits "send" and the time that a message leaves his in-box. He called this his "rage filter." He experimented with a couple of durations and found that the one-minute delay was just right. The sixty-second countdown provides a bit of drama and motivation to catch a problematic message in time. On a number of occasions, he has retrieved a potentially damaging message just seconds before it would have landed a bomb in a colleague's in-box. The short time window keeps him in the flow of dialogue at work in a way that a ten-minute delay (or other strategies such as removing the recipient's address until a message was perfected) would not. Occasionally he has used the delay to sharpen his message and express urgency, but typically he reworks the messages to soften the tone to remove any note of personal accusation. This delay generally suffices. For the most sensitive communication, though, he draws on another device, the "drafts" folder.

He recalled a confrontational note he wrote to his father when his mother was in the late stages of cancer treatment. As Matthew was taking his mother to dinner, his father warned them about the many things that his mother should not be eating. In an email afterward, Matthew explained to his father how absurd and petty those warnings were given that this could be the last meal that he and his mother enjoyed together. He realized later that his father was overwhelmed and simply trying, albeit in misdirected ways, to care for his wife. The email remained in his drafts folder. Some messages, he has learned, should never be sent.

A variety of tools have surfaced to serve similar functions as Matthew's rage filter and drafts folder. Consumer apps offer tips on writing emails in a way that will elicit a response or will appeal to the recipient's personality.[13] And researchers have offered feedback for politeness and rhetorical strategies within group forums.[14] My colleagues and I developed a prototype system to give feedback about emotional mirroring within email, building on decades of research in sentiment analysis and language style matching by James Pennebaker and colleagues.[15] Comprehensive analytics—on indicators of personality traits as well as emotional tone—can be gleamed from any sample of text that is entered into the Tone Analyzer from IBM Watson, the artificial intelligence (AI) system that famously beat the best human players at *Jeopardy*.[16]

Despite such advances, and extensive research in natural language processing and computational linguistics, feedback to help individuals improve communication is not yet fully integrated into everyday communication tools.[17] For the time being, people like Matthew will be left to create systems of their own. This is not unfortunate. In addition to the help that they receive from these systems, the mere act of creating them cultivates agency and self-efficacy.

Dating by Committee

Sometimes it helps to see ourselves through the eyes of an outside observer. Therapy and coaching can help us become more aware of ingrained patterns of behavior that we would like to change, but it is hard to catch ourselves doing these things in daily life. We may not realize that we are crossing our arms, mumbling, or dominating the conversation. And even if we do catch ourselves, it can be hard to quickly find an alternative stance or tone. In such situations, real-time consultation with peers could nudge us to change our behavior on the spot.

One artist and developer, Lauren McCarthy, experimented with a new way of getting real-time coaching. When she went out with men she'd met on OkCupid, she brought two phones with her: one livestreamed the date conversation to Mechanical Turk workers, and the other phone streamed their observations and suggestions back to her.[18] Mechanical Turk is Amazon's employment marketplace, typically used for crowdsourcing menial tasks at a low cost, but Lauren turned it into a tool for interpersonal coaching. Lauren's coach could be anyone who signed up for the task, but one can imagine a future in which workers are rated based on their interpersonal advice along with the other criteria typically valued in crowdsourcing, such as efficiency and accuracy.

Lauren's Mechanical Turk coaches said things like "Lean in," "Ask him about his favorite movie," and "Tell a joke." Carrying out some of these directives and acting out of character was freeing for her. The dates tended to go fine, so she realized that she had more choices about what she might say or do than she would otherwise presume. Before this she felt bound within tight parameters, as if there were only a few acceptable things to say in any situation. The feedback was occasionally insightful, but it primarily had value simply because it was from someone else who was watching over her. The messages made her feel like she had wingmen, that she was not alone. Walking home after one date, still logged into the app, she felt comforted by the idea that one of her coaches could speak up if she ran into trouble.

Subsequently, many others have used the streaming app that emerged from Lauren's early experimentation, Crowdpilot.[19] It provides real-time coaching not just from Mechanical Turkers but also from Facebook friends. Individuals can invite specific friends to listen into a conversation and specify what type of help they want. Some women in India have found it valuable in a particularly high-stakes situation: a date with a marriage candidate selected by one's family. This single meeting can determine whether a couple will proceed with an arranged marriage. Women have used the app to bring their friends virtually along on that date, getting advice about what questions to ask and opinions about whether they should go through with the marriage. While there is not extensive feedback from the women, several said that it helped them feel confident about their decision to either accept or decline a proposal. As Lauren experienced after her own experimentation with dating by committee, however, there was a downside: these women described feeling uncertain about how to act in subsequent interactions without the app running.

Dating consultation is the tip of the iceberg for how we might be using the capabilities of our phones for streaming given the challenging conversations that occur at work, at home, in the doctor's office, and other contexts. Whether we are negotiating household chores or writing to a new love interest, chances are that there is a friend or acquaintance with relevant expertise. Ideally, this streaming, when combined with social networking profiles, will help us tap peers who are especially knowledgeable about a specific topic rather than rely on one friend for all predicaments. And when anonymity is important, Mechanical Turk or a peer group outside one's social network may be helpful.

People have long sought peer advice through other communication technologies. It's common to submit draft emails for a friend's approval before they go out to the intended recipient or share pictures of outfit options for input. This consultation often runs parallel to other communication: one student described how her extended FaceTime sessions with a friend are punctuated with Snapchat captures of online purchases that her friend is considering. It is more challenging, though, to elicit real-time feedback about spoken conversation. There are now several streaming services akin to what Lauren hacked, such as Periscope, Facebook Live, and Instagram Live. These don't yet have consultative features, but it wouldn't be surprising if some individuals were informally receiving that coaching from their followers.

Along with opportunities for such support and feedback, Lauren's provocative use of the phone brings up obvious privacy concerns. Sensitive information about other people can be exposed whenever we seek coaching about that interaction, especially when images, video, or audio are relayed. Some people try to blur images or crop out identifying information before attaching or posting a screenshot, but it may still be possible to infer identities. The person who is consulted may know both parties and may pass the screenshot along to others or perhaps even post it publicly. These kinds of worst-case scenarios should be weighed before sharing a message that someone else believes is private. The app that evolved from Lauren's early experimentation, Crowdpilot, does instruct users to inform conversation partners about streaming, but this is far from a guarantee that such disclosure will take place. It's conceivable that apps could require spoken consent and voice-based verification from anyone who is being recorded. Ultimately, however, the responsibility rests with the user to apply such technology thoughtfully.

Extensive back channeling with an adviser can of course interfere with the conversation for which consultation was sought. This was the case for Taeyoon, whose blind date was also using the app (as part of design research conducted by the developers).[20] Bombarded by the feedback on their phones, he and his date barely engaged with each other. Their conversation felt hyperstimulated and comically disjointed. It was only after both of them turned the app off that he felt any sense of connection. But by then, it was apparently too late: his date wandered home, rejecting his invitation to a show after dinner.

Perhaps a bigger risk is following the counsel of a remote adviser who is missing out on important cues. Not long after the release of Crowdpilot, such a scenario played out in the "Christmas Special" episode of the British dystopian science fiction series *Black Mirror*: a seduction expert remotely observed his client, Harry, through an implanted augmented reality device, called "Z-Eye," that captured Harry's field of vision and displayed it on the screens of the expert and other observers. Through the device, which also transmitted voice, the expert encouraged Harry to approach a particular woman at the party, and later on to go home with her. Harry voiced his concerns to the expert through the device when this woman stepped away, but he ended up following the advice and leaving the party with this woman. Because he was only seeing things from Harry's point of view, the expert missed a key part of the interaction: the woman struggled with auditory hallucinations, and after watching what appeared to be Harry arguing aloud with himself in the midst of a crowded party (as he argued with the expert through the device), assumed that he was similarly tormented. Later, she poisoned herself and Harry.

Most streaming consultation won't end in this kind of Shakespearean drama, but the episode highlights the importance of getting perspectives beyond one's own. Even expert advisers may be missing crucial cues. With that caveat in mind, real-time consultation may be a powerful avenue for giving and receiving support.

"I'm Your Stress Phone!"

When a child gets in trouble at school, a parent may feel badly about everyone involved. Disappointment, embarrassment, and sometimes protective anger can all leak into a discussion with the child, exacerbating conflict. I've been impressed by parents who collaboratively work through conflict with their kids, especially those who go as far as sharing their own self-help tools.

Eliza, the mother of a seven and nine year old, juggles parenting with a demanding full-time job, expectations of her extended family and a passion for long-distance running. She struggles to keep everything in balance and hates feeling like she has let anyone down. This happened when she got a call one afternoon from her older son Nigel's school, with the complaint that he had disrupted chess class. In her words, she "blew up" at him when he got home from school. Later that day, she sat down to talk with Nigel and shared the Mood Phone, which she had been using for managing stress and anger. The app included various ways for people to indicate their current emotional state (such as the Mood Map described above and scales based on visual metaphors such as fire) along with cues for reframing, perspective taking, and problem solving.

Eliza and Nigel used the app as they talked through the entire incident, touching the images of fire that represented stages of anger, and exploring the cues for reinterpretation and constructive communication. They started to see the incident from one another's perspectives: Nigel said that he and his friend were just having fun by tying their shoes together, Eliza acknowledged that she was embarrassed to get a call from the school, and the teacher, they guessed, was probably just trying to keep chess class under control. Their joint exploration helped them let go of inaccurate attributions, such as "the teacher hates me" and "I'm a terrible parent," and acknowledge that the rage they felt earlier had subsided.

It is possible that the visual metaphors of fire helped Eliza and her son acknowledge their anger. Building on Lisa Heschong's description of fire, "It could grow and move seemingly by its own will; and it could exhaust itself and die," perhaps Eliza and her son were able to see, in the images of fire, the vitality of their own anger along with its phases and impermanence.[21] They could see that they were safely in the aftermath of anger and recognize future opportunities to catch the flicker before the flame.

Eliza notably shared her own self-help tools—in this case, a mood app—with her son. And she didn't just hand off her phone to him; she incorporated the app into their conversation. As a result of these conversations, their communication improved, and some of her parental stress decreased.

They interacted with the phone in a different way than they would have treated a book or other analogue tool. It became a part of their conversations and how they talked about their relationship: Nigel sometimes reminded Eliza to bring her stress phone as she was leaving home and started coaching her with tips from the phone, or what he thought the phone might say in a specific situation. They engaged with it almost as if it were alive. On one occasion, her son joked, "I am your stress phone!" Identifying with the phone may have helped him feel closer and more helpful to his mom. What started as a tool for Eliza to manage her stress individually became instead a medium for talking with her son. With it, she addressed the source of her stress, not just her reactions to it.

Anxiety Is Child's Play

Eliza turned a self-help app, designed for her exclusive use, into a conversational aid that she shared with her son. With the app, they were able to talk about the fight that they had earlier in the day and understand each other's feelings. In the next example, Silver's son pushed her to share this same personal stress management app. Here, the sharing took the form of play. Silver's son was too young for a full conversation, but by using the app with him, she improved their connection. Both Silver and Eliza ultimately reduced their own stress by calming their children.[22]

Silver was chronically anxious, in large part due to her son's health problems. A single mom with a stressful job, she never felt that she was on top of things. She rushed her son out of the house each morning, regretting that she couldn't enjoy his curiosity and emerging preferences: "So those are all normal three-year-old behaviors. ... I can't relax ... and enjoy watching him decide which coat to pick out or, you know, whether to wear the Spiderman shoes or the boots with the dinosaurs. ... Those should be positive things in life."

She used the Mood Phone app (the same prototype used by Eliza and Tobias above) for monitoring her mood and practicing stress reduction exercises for one month. Once, when her son was upset at a dental appointment, she took out her Mood Phone and did the breathing exercise with him. Afterward, he kept asking to do the breathing with her again. "Like even this morning, my phone went off and he wanted to do breathing. ... 'Please show me the breathing.' ... Or 'Let's do breathing.'"

Her son clearly wanted to do the breathing exercise with her even though she had initially given it to him as something he could do on his own. Like many busy parents, Silver sometimes needed to distract her child. But at that moment, she also saw the option of using the technology as a way to play with him. This was easy because her son immediately experienced the breathing exercise not as mindfulness training but as a game.

In one interview, Silver described a dream in which her Mood Phone had been taken hostage and she was able to see, through a second phone, some of her self-tracking data. It was as if she was watching herself being taken hostage by her anxiety and looking for an escape.

There is value in examining our attachment to particular technologies. As Sherry Turkle shows in *Evocative Objects*, cherished objects come to represent how we see ourselves, our defining relationships, and our choices.[23] In the dream recounted above, Silver saw how she was bound by anxiety and unable to experience small joys as a result. The notion of being so attached to our phones that we dream about them may strike some people as unfortunate. But Silver used the Mood Phone as a transitional object, a device for psychological growth. The transitional object is a concept developed by British psychoanalyst D. W. Winnicott to describe a familiar possession, prototypically a favorite blanket or stuffed animal, clutched by a child to explore the world apart from the familiar parent.[24] It is not just in infancy that we draw on transitional objects, though. Silver became attached to the phone as she transitioned to a broader range of emotions and ways of relating. She trusted the Mood Phone with her feelings and was reluctant to return the phone at the end of the study. That attachment to the device paralleled her anxiety and the feeling that she needed to be worried all the time. Eventually, she felt OK about detaching from the phone and her anxiety. She became interested in relating more freely with her son in the moments that she didn't need to be acutely concerned about his health.

Silver's experience bears some resemblance to the attachment depicted in Spike Jonze's 2014 film, *Her*. It was only after falling in and out of love with his operating system that the main character could shift his attention back into human relationships. We should take our attachment to technology seriously—precisely so that we can get on with our interpersonal relationships.

Family Phone Contract

Concern about strong emotional attachment to technology made one father hesitate to get his teenage children phones of their own. Peter and his wife live in Barcelona with their two children. Like many parents, Peter worried about his kids and technology. In addition to the financial burden of giving them phones, he worried about the toll on their social skills. He dreaded the idea of his kids staring at their phones during dinner or playing mobile games when they should be doing homework. While visiting the United States, he attended a talk by Sherry Turkle, who articulated his worries with examples of teens who avoided spontaneous conversation, and whose social media use left them feeling inadequate and alienated.[25] But increasing pressure from his children recently made him wonder if his previous "just say no" policy was tenable.

Peter's son, Davide, complained that he was being excluded from events at school because he did not have a phone. All the other kids on his rowing team had their own phones and coordinated their practice times and rides using WhatsApp. His mother joined the WhatsApp rowing group to ease coordination. This was logistically helpful, but Davide was humiliated when his teammates puzzled over why his mother was messaging them. At this same time, Peter's daughter, Pamela, pined to join her friends on Instagram, Snapchat, and other social media. His wife's phone, it turned out, needed upgrading. These demands were all the more stressful because he had recently transitioned from a big company job to a less certain start-up.

Peter stepped back to think about how he could manage the interpersonal and financial costs associated with phones. In the course of talking with friends, he saw that he had some options. He decided to write a contract that explicitly outlined the terms he hoped were already understood: the phone was not to be used at meals, brought to bed, or used to extend the time that his children were given to play games each day. It satisfied Peter and his children. His kids were pleased because they got phones and were not surprised by the conditions. For Peter, the contract provided reassurance that he would not lose his children to technology.

Within their small community, this contract went viral. His daughter mentioned it to a friend at school, and word got around. Other parents texted Peter, including ones he had never met, asking for it. He shared it with them as a Google Doc. Once the family agreed on the terms of the contract, Peter and his wife passed down their out-of-date phones to their kids, thereby managing both financial and interpersonal concerns.

Peter took the concept of a phone contract, normally an agreement between an individual and large company, to establish norms within the family. This contract allowed him and his wife to spell out their concerns and allowed his children to discuss them and request any necessary modifications. This approach gave him a feeling of control over what had previously felt like a dangerous and slippery slope.

Devices as Bridges

The men, women, and children in the stories above used their devices to stimulate conversation and foster intimacy. In some cases, devices that were intended for personal use were shared: a man used errant scale data to start a conversation with his friend, a husband drew on his mood app to empathize with his wife, and a mother and son were able to take each other's perspectives by sharing a stress app meant for her private use. In other cases, social media was used to address specific individual needs. For example, it was through an asynchronous word game that a socially anxious family member reentered family conversation. In the last story, the mobile device that would help an adolescent's friendships threatened family relations. His father could have used an app that curtailed his children's use of the phone. Instead, he deployed a purely social device, a contract, to limit those risks and open up a conversation about the terms of use. Throughout these examples, individuals found intimacy by focusing on their relationships and goals rather than on the technical device. They understood that the device was merely a bridge, not to be mistaken for the connection itself.

3 Meaningful Measures

We are, by default, tracked. We are assessed by our own devices and the devices that surround us. The challenge is to derive meaning from these measures.

Self-tracking ranges from casual accounting of health behaviors or productivity, which has become the norm, to highly motivated self-investigation.[1] The Quantified Self (QS) movement, started in 2008 by Gary Wolf and Kevin Kelly, brought together these motivated self-trackers. In QS meetups, which continue today around the world, individuals share their tracking experiments: their objective, what they measured, and what they learned. The anticipation of presenting to a QS meetup often intensifies this investigation and audiences typically respond in a supportive and collaborative style, indicating that despite the individualistic focus suggested by its name, the QS movement at its core is social.

Early QS investigators often created their own tracking systems to conduct their experiments. The late Seth Roberts, professor at the University of California at Berkeley, for example, wrote a program to test his cognitive functioning (it presented simple math challenges on his computer throughout the day).[2] He found that his arithmetic was much better when he ate large quantities of butter, extending his previous personal experiments on fat consumption and cognitive performance. Many other QS experimenters have similarly examined how their productivity and well-being varied as a function of dietary choices, sleep, and other behaviors.[3] Kevin Kelly, in a keynote address at the 2012 QS conference, explained how tracking and feedback are expanding the self. With extended sensing capabilities—"exosenses"—and AI, Kelly argued, we will not pour over numbers to learn; instead, we will immediately feel the data collected from our own sensors and those of other self-trackers.[4]

We can see examples of this enhanced sensing. Take the personal experience of Michael Snyder, genetics professor and precision medicine researcher at Stanford University. On a flight to Norway in 2015, wearing eight commercially available biosensors, he noticed that his heart rate increased and his blood oxygen levels dropped (as was typical for him during flights) but did not normalize on landing. These measures, along with other symptoms and his recent visit to rural Massachusetts, raised his suspicions that he had contracted Lyme disease. He convinced a doctor to prescribe a course of antibiotics. The Lyme organism was picked up in subsequent blood tests, confirming that this was the right course of action.[5] Had he not drawn meaning from his sensor data and acted quickly, the treatment might not have been effective. In Lyme disease and many other conditions, early diagnosis and treatment are critical.

Such systematic self-tracking practices have helped establish precedents for collaborative diagnosis and given rise to a science of single-case designs that may help those with rare diseases and the many others who are trying to track down their particular triggers for more common conditions.[6] For some, like Jennifer Mankoff, a computer science professor at the University of Washington who also suffers from Lyme disease, tracking informs difficult treatment decisions. She had started a difficult course of treatment and didn't know whether to continue it. She had so many symptoms and was in so much discomfort that it was hard to assess progress. Through detailed tracking and a regression-based statistical analysis, she realized that although her symptoms were not decreasing in intensity, they were decreasing in frequency. With this numerical insight, she was able to see concrete evidence that the treatment she had started was effective and worth continuing.[7]

While early QSers had to hack their own self-tracking tools, now we have to make special effort if we don't want our every step and heartbeat recorded. Our phones and wearables capture and also emit data about us constantly. Some websites track microbehaviors, such as changes in facial expressions, gaze, and mouse movements. Businesses, governments, and friends track us.[8] Some cars attempt to assess cognitive and emotional states, an effort that will advance as research on affective computing is implemented.[9] Our homes will become increasingly "smart," generating more data from wirelessly connected thermostats (such as Google's self-adjusting, remotely accessible Nest), lights, locks, plant sensors, ovens, refrigerators, and myriad other devices.[10] Meaningful illustrations of our daily routines may emerge as data is gathered and integrated from these various devices.

Although this nearly ubiquitous sensing is rich with potentially helpful information, it is not always to the benefit of the self-tracker.[11] Those who use self-tracking devices are typically sharing their data with industry, where it may be used for customer profiling or algorithmic development. Tracking is bound up in the applications that we have come to rely on in daily life (such as maps and ride sharing). And using GPS, a free government service, often means allowing tracking by the companies providing these applications. That we are not the only ones looking at our data is all the more reason to try to get something back. Many tracking devices now attempt to go beyond sharing insights from data to intervening on our behalf, notes cultural anthropologist and media scholar Natasha Dow Schüll.[12] As an example of what she sees as a kind of infantilization, Schüll points to HAPIfork, which vibrates if one is eating too quickly, essentially enforcing a slowdown.[13] Rather than off-loading self-regulation, Schüll sees the strength of QS in terms of explorations in sensing and feeling.[14]

To ensure we benefit from the tracking devices that may be in, on, and around us, it helps to get clear on what we are seeking. In the stories below, individuals measure (or in some cases, decline to measure) according to their own objectives. Most were trying to bring about a change in their behavior or outlook, or sustain a change. Often that required them to reject or rethink the persuasive messaging from their devices. Part of the power in their strategies is that they were self-improvised rather than prescribed. Individuals who want to increase their step count, drink more water, or progress through their to-do lists more efficiently will probably be better off using products that were informed by principles of persuasive design than those without such a basis.[15] But the goals and challenges faced by the people in this chapter were not completely aligned with those presumed by their technologies. Some appropriation, or hacking, was required.

Their stories are intended as an invitation to measure with meaning. They also point out the importance of considering how to involve others. For, even though we typically have little control over the distribution of our personal data, sharing within our relationships and communities is something we can thoughtfully determine.

No Size Fits All

Kaiton has his own style and rarely uses anything out of the box.[16] He tailors his T-shirts and repurposes furniture. When he set out to lose weight, he took a similar approach of sampling and tailoring.

Kaiton wanted a simple and accurate way to track what he was eating and systematically evaluate what type of diet was best for him. After trying a number of apps, he realized that some of the ones with the best tracking features unfortunately also had preprogrammed motivational messaging. He couldn't relate to this approach at all; he's just not into autogenerated congratulatory emails and badges. So he filtered these emails and trained himself to ignore the sections of the screen about badges and comparisons to other users of the apps. He also had to put effort into translating his food preferences, which emerged from his health interests and Jamaican farming heritage, to align with diet-tracking features that primarily accommodate prepackaged foods and items from chain restaurants. Through dutiful tracking, he found that a protein- and vegetable-oriented diet, involving intermittent fasting, felt right for him. His personal discovery, of course, echoes well-publicized research. Even though it may have been headlines that led him to test out these particular practices, and while he may have received exactly these same recommendations from a health provider, his personal discovery process engendered a sense of control.

Kaiton lost thirty pounds over the course of two years. During this time, he monitored himself with myriad apps, wearable devices, and periodic physiological tests to determine body composition. He tracked his diet and exercise by cobbling together features of different applications, toggling between diet- and fitness-tracking apps, such as Lose It! and Moves, and mapping visualizations. And at no point did he lose his skepticism about motivational messaging or references to a community of fellow self-monitors. These rewards had no resonance for him. His ideas of community ran deeper than a user group who had downloaded a particular app. Kaiton was successful because he was confident that he could adapt technology, just as he tailors his clothing. He assumed, correctly, that he could meld the appealing components of different systems and reject the rest.

His success also depended on his self-awareness. He became fascinated with how the measurement process was consuming him: he noticed that he was starting to let the data define him and that he was obsessing over every gram he consumed. Kaiton acknowledged that he became tyrannized by numbers, and that he was almost always juggling calculations of calories and nutrients—not just when he was eating. He reported his diet strictly, sometimes overestimating what he ate to ensure he'd get tough coaching rather than cheerful reassurance from the apps. Even while in the throes of this obsession, though, he was mindful of it. Ultimately, it was in accepting the power that this data had over him that he found some freedom. As his tracking continued, he was able to maintain his progress while broadening his perspective.

By reflecting on his self-tracking, Kaiton ultimately moved beyond increasing his control over fundamentals of nutrient intake and expenditure. As he examined the quantified measures of paths that he had walked on a given day, particularly while traveling, he found himself imagining the paths not taken; he wondered about what might have happened one day, for example, had he walked up the hilly street from Taksim Square instead of taking a taxi. Beyond the numbers, what might he have seen? And this led to speculation about other cities that he could walk through and friends he could meet on his explorations.

Over the course of his tracking, his questions broadened from body composition to life composition. The counting of steps, calories, and nutrient milligrams could have led to a rigid lifestyle governed by strict policies about intake and exertion. Instead, he broadened his focus. With the mash-up of data, he pushed himself not just physically but also personally to imagine future paths that could enrich his life.

Greens for Likes

Others are following Kaiton's approach, turning toward qualitative tracking and forming their own online communities. Image-based social applications, such as Pinterest, Facebook, Instagram, and Snapchat, have always been used to broadcast a delicious meal, mastery of a complex recipe, or a visit to a special restaurant. Increasingly, people are also posting to find support for eating healthfully.[17]

One woman interviewed by Christina Chung, for her dissertation research at the University of Washington, started eating more greens after realizing that they elicited more "likes" than other dishes. Salads and other vegetables photographed better than rice, for example. Her shift toward vegetarianism was loosely consistent with her health goals, but it was clearly instigated by this social response. You might think that she would continue eating as she had been but only post photos of greens. This woman and others, however, felt compelled to document absolutely everything they ate in order to be truly accountable and did this with ease by snapping photos.

While diet-tracking apps are tedious to use and easy to forget, many people already use Instagram to capture what they eat and reflect on their data—as they browse through the images that they've posted for likes and comments. As Chung and her colleagues found, some people appreciate that Instagram is not a health tool and does not have a space for detailing specific nutrition data. Rather than selecting approximations from a checklist, they can show their actual meals along with information such as portion size that might otherwise be difficult to convey. With hashtags, they shape their support communities, borrowing from the terms that they see in related posts. In the course of seeking support for changing their own diets, some of the individuals in this study found a mission of inspiring others and continued posting fastidiously even after reaching their personal goals.

Not Counting

When Masie upgraded her iPhone several years ago, she encountered an unexpected feature: it automatically gathered data about her activity from the embedded Health app. She had not requested this data and didn't want it.

For years, she had struggled to turn off her own mental clocking of physical activity and logging of everything she ate. Masie had been anorexic as a teenager, and in recovery she learned to let go of counting and focus on the intrinsic pleasure of movement. As an adult, she developed a love for walking and dancing—pleasures she didn't want contaminated by obsessive thoughts. So for her, the activity data on her iPhone was a step backward, triggering her latent compulsion to over-exercise. Although fully recovered, she always experienced the step count as an admonishment to walk more. That undermined the joy of walking and her intuitive sense of how much movement was right for her.

It felt like the data was forced on her. It wasn't just that she had never installed the application; she couldn't even uninstall it. Neither could she disable the sensor tracking of her movement or frequent updates. To block it from her view, she hid the app within the utilities folder. Even so, she had to regularly delete the data to prevent it from appearing on her Health app dashboard. Her iPhone really, really wanted her to know the activity data it was gathering and apparently also to think about this data in the way that Apple thought everyone should—that is, as encouragement to walk more.

Masie knew that she could cope with this, but she was concerned about the effects of this feedback on others who were prone to compulsive exercising. As a health activist who reaches a sizable audience on social media, she used Twitter to petition Apple to make the feature optional. She doesn't take credit, but Apple eventually shifted its policy to make the count optional. Under privacy settings, users can now turn off motion tracking and delete any already-collected data within the Health app.

Things could have played out differently. Masie could have responded to the step count data by relapsing into a self-focused perfectionism. Like Kaiton, however, she decided to take control of the data being presented to her. She did so as an activist, enlisting her professional and social contacts to join her mission. Instead of collapsing inward, she reached outward to change a product that affects vast numbers of people. Seizing control over choices that Apple had withheld from her led to an increased confidence. She saw that she could bring about change.

Data is supposed to be the bedrock of truth, akin to reality itself. You can't argue with truth! But data is often presented back to us in ways that reflect designers' assumptions about our motivational structures. One assumption is that all users need to be pushed into exercising more. Of course this will not work for everyone, and it may be dangerous for some. It is not necessarily the case that data will motivate us to make healthy choices. And it is certainly not true that the same choices are healthy for everyone. Even something as seemingly benign as increased walking can function in different ways. Walking is, in general, a helpful form of exercise, but not if it plays into procrastination or compulsion. Self-monitoring of movement and eating can be counterproductive among people who are already restricting their diet and preoccupied with exercise; it can trigger symptoms in those who are vulnerable to disordered eating or exacerbate an existing disorder.[18] One-size-fits-all feedback can also be irritating for those with physical restrictions (imagine a woman in a wheelchair being prompted to take more steps). There has been some effective resistance to products that present generic nudges regarding diet and exercise. For example, in response to customer complaints, Google Maps stopped its brief experiment of displaying, through mini-cupcake icons, the number of calories that a user would burn by walking different routes.[19] But the marketing of most wearables and self-tracking apps implies that the same information will have the same positive effect on everyone.

As Kaiton and Masie have shown, we can disagree. Pushing back against these assumptions can have its own rewards.

Blood Relations

Jessica was diagnosed with diabetes at the age of four and endured a childhood of injections, monitoring, and close regulation of her diet. Her condition and the rigid way that insulin formulas worked at that time set her apart at school, sometimes causing her to feel isolated. The other kids in her fourth-grade class, for instance, saw her scheduled snacks (needed to avoid hypoglycemia) as a luxury. "It's not fair," they complained. The teacher fortunately had the great insight to make it snack time for everyone—a move toward inclusion that Jessica still appreciates decades later. She felt part of the group again.

But as she got older, social pressures became more complicated. In her early teens, she had friends who began to experiment with ways to lose weight—exercise, restricted diet, diet pills, and so on. This affiliation had benefits beyond companionship: since these friends also had a complicated relationship with food, there wasn't social pressure to eat. It was clearly not a healthy route, but such limited eating made it feel easier to control her blood sugar and reduced her need for injections, which she was embarrassed to do in front of friends. In her later teens, she tired of this excessive control and her near-lifetime focus on numbers. She then swung in the other direction. In an attempt to feel normal, she stopped monitoring her blood sugar and stopped exercising. This phase was far from carefree, though. In the back of her mind, she worried about the damage she was doing to her health.

In her early twenties, she saw a new endocrinologist who helped get her back on track and urged her to use a pump to deliver insulin rather than injections. Even though she still had to check her blood sugar frequently during the day, the pump provided more flexibility, and she found it much easier to deliver insulin by pushing buttons than by injecting. The pump presented its own set of issues, though: the device looked medical, and she struggled with how to wear it twenty-four hours a day.

Jessica was finishing up her undergraduate degree in digital media at the time and was driven to do something about the design of the insulin pump. She began a graduate program where she could focus on the design of health technology and took this on as a challenge. As part of her research, she sought out others living with type 1 diabetes. They echoed Jessica's own complaints about the pump. It interfered with the line of any outfit, whether a dress or yoga wear. And it interfered with their identity and sexuality, whether they were wearing clothing or not. The device signaled illness and disability. As Jessica developed new ideas for clothing that addressed these concerns, she bounced them off women in her support group. The designs that emerged from all this, Hanky Pancreas became popular. She later broadened her focus from designing clothing accessories for a specific disease to incorporating the patient's experience into many types of health care design.

Several weeks before our first conversation, Jessica started using a continuous glucose monitor. The device, self-inserted every week, transmits updates of her glucose-level readings to her iPhone and Apple Watch every five minutes. It allows her to share readings with others, who can get customized alerts on their phones. The sheer novelty made it fun to share readings with her mother, boyfriend, and sister. And the sharing didn't stop there: one night, her sister showed it to a friend who was visiting, and, as if they were watching a reality TV show that involved viewer voting, she and her friend offered their thoughts. (See screenshot of their exchange on the next page.)

There is a cascade of meaning in this simple exchange. If the continuous glucose monitor did its job without the functionality of sharing and pushing notifications, it would have served its purpose for Jessica: to issue a stream of numbers of vital significance to her, the person living with diabetes. But the sharing of these vital numbers enabled important connection with her family. They now had a reference point for expressing concern and were able to show support more frequently.

This has been Jessica's story from the beginning. Through devices intended to support pressing biological needs, she has established meaningful connections. She's moved away from feeling isolated by the disease as she did as a kid. She uses her current devices, the insulin pump and continuous monitor, to connect rather than hide her condition. Through

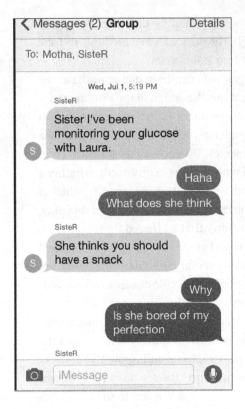
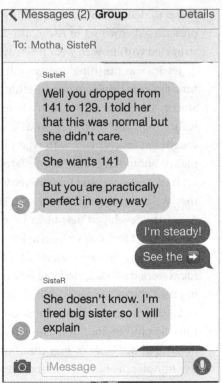

these devices and those that she's created, she's found greater purpose and involvement. Her engagement in a large Facebook group of patients who use these devices, online conversations with a smaller group of women with type 1 diabetes who she has known for years, and the close sharing of data with her boyfriend and family each fosters fosters a crucial and different kind of connection.

The Highest Form of Flattery

Self-tracking isn't limited to carbohydrate and step counting, of course. Amit, a graduate student in engineering, used a language-tracking app to help him with his communication style. English is not Amit's first language. He had moved to the United States to start graduate school a year before we met. While he struggled to find the right way to express himself, he admired the way that his classmate and friend, Amanda, communicated. Even in email, Amanda came across as both warm and serious. He noted that she always started her messages by thanking other members of a project team for their contributions, but he also knew there were subtleties to her style that added to its effectiveness.

Amit started using a program my colleagues and I developed that scored how well an email response matches the tone and style of the original message. We designed it to help people get "in sync" with others by mirroring their language.[20] Language style matching has been associated with rapport, and some research suggests that when people mirror a conversation partner, it builds affinity and positive outcomes in negotiation.[21] The tool that we developed displayed the similarity between the two messages in emotional tone and other aspects of writing.[22] As someone composed a message, the system provided matching scores as feedback, dynamically updating the scores as the message was edited.

Amit told us that he didn't always know what to make of the feedback that the system was giving him, but he had a more specific aim in any case: he didn't want to learn how to mirror everyone so much as to learn how to become like one particular person—Amanda—whose social adroitness he so admired. So when he was responding to Amanda's emails, he tuned his writing to match her style, using the mirroring index as a guide. From this he learned to borrow Amanda's phrases of speech, structures of sentences, salutations, and greetings. Subtle "function words" like pronouns, included in the feedback metrics, changed along the way. He started writing this way not just to Amanda but to everyone. Amit successfully turned our general-purpose mirroring tool into a tool for style emulation.

Online and off-line, we are influenced by the communication styles of those around us. By adapting a tool without much concern for what it was supposed to do, Amit was able to heighten this influence and tune his communication in precisely the way that he sought. In restyling his written voice, he was reconstructing his social presence.

Family Planets

In tracking another person, we sometimes learn about ourselves. Reflecting on data about her mother's social isolation, Natasha saw a need to change her own life.

Sandwiched between the needs of her eighty-five-year-old mother, her children, and their children, Natasha felt depleted. She was mostly worried

Social activity tracking. A conceptual sketch of the feedback for older adults
and caregivers.

about her mother, Mallory, who she knew must be lonely. Mallory enjoyed
spending time with others and seemed to want more contact, especially
with Natasha and her family, but she was never the one to make it happen.
Her husband had been the social coordinator in their relationship, and her
friends were people they had socialized with as a couple. Mallory moved
across the country to live near Natasha after her husband died, becoming
more cut off from her few friends and extremely reliant on Natasha. Unless
Natasha arranged it, her mother saw no one. This passivity was not just a
burden for Natasha; it also concerned her. Although she had some worries about her mom's physical health, more than anything she wanted her
mom to be actively engaged with other people. Her concerns are consistent
with growing evidence that loneliness is a serious matter, reducing life satisfaction and increasing the risk of cardiovascular disease, dementia, and

other illnesses.[23] And it's not just being around people that would help someone like Mallory. To feel connected, she would need to reach out to others and participate.

A project that my colleagues and I had been working on fit the bill. We had designed a system that tracked as much of an individual's social activity as it could, including face-to-face interactions, phone calls, and emails. It then presented that data as a solar system, with people shown as planets circling the elder in the center.[24] The distance of each planet from the sun represented how much contact each family member had been having with the older adult. It was technically challenging to gather the data. But the visual representation of that data as a solar system turned out to be what mattered, in some unexpected ways.

The solar system metaphor allowed Natasha to monitor and make sense of her mother's social situation, moment by moment, over the months that they used this system. Perhaps more valuable was that it gave her a vocabulary and external reference point to talk with her mother about loneliness. It showed her mother as a circle at some distance from others, except Natasha. Natasha discussed this objective data, sometimes using metaphor, as a way of sharing her concerns. And she drew on the data in our interviews to sympathize with her mom's isolation and even her reluctance to invest in new friendships. It must be, she said, "like being on an island, when everyone else you've known and loved has died." She struggled to imagine how hard this was.

Unexpectedly, Natasha also used this solar display as a sort of mirror rather than just a way to check in on her mother. She saw that week after week, she remained at the center of her mother's social life. Hers was the only circle close to her mother's. Her central position in the display of her mother's world reflected her dutiful caregiving and hinted at all that she had been neglecting. Up until then, it was as if she had been so close to the center that she was blinded. By stepping back and seeing her situation objectively, she could start to examine the rest of her life. She resolved to draw in other family members as caregivers for her mom so that she could allocate more of her time to her other relationships and interests.

Data by itself does not tell the story. Natasha had to see that data expressed in terms that made visually clear the extent to which she was at the center of her mother's universe. The metaphorical display let Natasha back up just enough to see that she needed to invite other "planets"—her siblings and children—to bring their orbits closer to her mother and help out. This shift in conversation and caregiving dynamics even sparked a change in her mother's passivity: Mallory started initiating plans with her family and even volunteering.

For data to make a difference in our lives, we have to make it our own. But for many of us, this involves seeing the data and ourselves as more than numbers. The image of planets alone in space, circling an isolated object, worked for Natasha and Mallory where an index or graph might not have.

Do You Know Where Your Children Are?

Knowing where someone is often provides great assurance. Location sharing promises this same piece of mind, but the monitoring of one person's location by another can threaten autonomy. In the next two stories, we see individuals navigating this tension. Karthik and his sister, who we met earlier, are in their early twenties. They live at home, and their parents still worry about them. Karthik's parents asked that he and his sister keep them informed about their evening plans. This is an implicitly agreed-on condition of living at home.

They often forgot to share their location. Perhaps his sister, like Karthik, was reluctant to give pinpoint information about her whereabouts to their parents. In an attempt to comply with the request, Karthik linked GPS updates from his and his sister's phones to his parents' lights at home. The lights changed depending on what region of London they were in at a given moment. He assigned a color to each geographic zone by creating a simple rule using IFTTT (If This Then That). The lights turned red if he was in East London, amber if he was in North or Central London, yellow if he was at the bus station near their home, and bright green if he was walking home from the bus station. This system built on something that he and his family were already trying to do and quickly became part of their lives.

Karthik's solution negotiated the needs of different parties. The children were comfortable disclosing their general locations to their parents, and the parents were satisfied knowing roughly how close they were to home. Divulging precise coordinates would have made the children uncomfortable, and that information wasn't even relevant to their parents. When information is playing a social role, disclosing just enough is often crucial.

It also helped that Karthik took an experimental approach, trying out several interventions with his family, each for a week. He had clear intentions for each intervention, and expectations of how his parents and sister would respond. Relationships are almost always affected by the information generated by technology, but it is hard to predict what these effects will be. Karthik's parents might have assumed that without precise information, they would be anxious, for example, only to discover that all they really needed was reassurance that their son was on his way home, or in this or that part of town, and the feeling of ongoing connection.

The End of the Affair

After George had an affair, his wife of fifteen years set new terms for their relationship.[25] One was that he would share his coordinates at all times using a GPS tracker. There would be no more wondering where he was. He agreed.

This changed his relationship with his phone as well as with his wife. While he was once glued to his phone, he frequently started "forgetting" it. One incident of forgetting happened on his birthday. His wife had planned a special evening to celebrate and asked him to be home by six. Perhaps because he wasn't feeling celebratory, or maybe as some form of passive retaliation, he lingered at work. He dawdled on his way home as well, making himself even later. His wife was perplexed. She checked her phone again to see his location. She sighed with relief at the indication that he was in the house, thinking that she must not have seen him come in. She went into their bedroom but found only his phone.

The birthday celebration did not go well.

George may have ditched his phone not because he was up to anything nefarious but simply because he didn't like the feeling that his wife was tracking his every move. Unlike Karthik, who devised a means of blurring his location in the story above, George saw only a binary choice: the location sharing could either be on or off. Because he had damaged his wife's trust, he knew that if he turned off his phone, she well might take that as an indication of infidelity.

Trust is actively negotiated in many relationships, whether those are with partners, families, or colleagues. And often there are ruptures, even if it is not due to something as blatant as an affair. George's wife reached to technology in the hope of preventing future betrayals. Location sharing may provide reassurance, but it is unlikely to ensure fidelity or trust on its own. And coercion seems bound to backfire—at least in romantic relationships. The solution for this kind of rupture has to go beyond technology. Among other things, we need a tolerance for uncertainty that our devices aren't likely to give us.

From a friend's Facebook page comes a variant with a twist ending:

> As many of you know, my kids have phones and those phones have trackers and I track my kids' locations at will. I'm a big fan of this and happy to argue its merits to anyone who finds it freaky. What you probably *don't* know is that the same app also lets my kids track me, too! And I've been busted by them more than vice versa. E.G. today at 3:08 I get a call from my daughter, wondering why I haven't left home yet to come pick her up (at 3:15). She's right! I'm busted! O, brave new world, that has such people in't.

Tracking others may become more complicated than it first seems. It often reveals as much about the observer as the observed.

Measure for Measure

Our lives have become increasingly trackable. Our calories, steps, blood oxygen, and "likes" are easily accessed. We know where our children and partners roam. As our everyday appliances—phones, cars, and homes—become increasingly smart, it may become harder to disentangle from our quantified selves should we want to. But as we've seen in these stories, we can find meaning, even diagnostic insight, from this data when we align it with our questions and priorities. And even in the tracking of our more mundane activities, with some thought and creativity, we might find opportunities for navigating communication, changing our habits, or resolving tensions with those closest to us.

4 Remembering and Forgetting

How can we shape our environments, both physical and digital, to focus our attention on what matters most to us? What kind of help do we want remembering some people and forgetting others? How actively do we want to architect the digital influences on our decisions? Below, individuals grapple with these questions. Some are honoring loved ones who have died, others struggling to move on after breakups, and several are trying to stick to the plans they've set for improving their health. To facilitate remembering and forgetting, these individuals creatively extend technology, increasing continuity in some cases and separateness in others.

An AI Afterlife

In the "Be Right Back" episode of *Black Mirror*, a British television series, a woman's fiancé dies suddenly. As a consolation gift, a friend signs her up for a service that allows her to continue conversations with him. Although creeped out by the idea, she uploads images and voice clips that she collected over the years so that the learning algorithms can generate new utterances—words that her fiancé never actually said but conceivably could. The woman becomes engrossed in the subsequent conversations with this shadow of the man she loved and even upgrades her service to include a robot body. She depends on this interaction but is so frustrated by the robot's passivity that, at one point, she leads him to jump off a cliff, lamenting that her fiancé would never follow such orders. The robot resists at this pivotal moment—a sign of life at last. She brings him home, and their relationship continues.

An AI developer who lost her close friend tried to do something similar. When her best friend, Roman, died in 2015, Eugenia Kuyda developed an

AI system to continue texting with a semblance of him.[1] Roman had not
been active on social media, so Eugenia gathered text messages from ten
of Roman's friends to form a model of his conversational style, adapting
neural net software that had been developed for her AI start-up. She and her
developers created a chatbot that incorporates aspects of Roman's language
and conversational style. Because this is real life with the real limitations
of technology and not a science fiction television series, the bot ended up
serving mostly as a listening device rather than a conversation generator. It
became a way for Eugenia to say the things to Roman that she wishes she
had said when he was alive. Reactions to this project among their friends
have run from intense caution about denying death, to disappointment with
the technical quality of the rushed product, to enthusiasm for grieving with
contemporary resources.

We are not far from such capabilities, until recently the realm of science
fiction, becoming a real part of our day-to-day lives. Our social media feeds
may increasingly feature selfies and dialogue with the deceased given the
emergence of apps designed for presence beyond the grave.[2] And what if the
voice from the smart speaker kitchen were not Amazon's Alexa or another
home assistant but that of a deceased grandmother? I could imagine asking
my grandmother, channeled through a smart speaker, to relay a story or
give me some advice. This could be comforting. But ordering Grandma to
turn off the lights would strike most of us as disrespectful. As the etiquette
and norms for such exchanges emerge, they will illustrate what we desire
from this kind of technology and how we imagine its role in remembering.

Eternal Conversations

There is no one way to mourn. There are no agreed-on coping strategies for
those who lose a loved one. These uncertainties are magnified when the
mourning takes a technologically mediated form.

On the third anniversary of her husband's death, Monique wrote him
a letter, which she posted on her own Facebook wall. In it, she told him
that she cherished the last moments they shared and described the surreal
experience that followed. She recounted the comforting conversations with
people who had suffered extreme loss along with the misguided attempts
of others to lift her spirits. She didn't want to be lifted out of grief. She speculated with stark honesty about the difficulty of imagining future joy.

After Monique posted this letter on Facebook, some of her neighbors texted her to express concern about her mental state. She had anticipated these reactions and had considered not posting the letter to forestall them. But she was resolute. To prevent herself from backing down, she drafted the letter the day before their actual anniversary.

The letter was in part a defiance of norms about death and bereavement. Loss and the personal evolution that follows, she feels, should be openly discussed. In writing to her husband, she pushed against the idea that she should treat death as erasure and the accompanying expectation that she should not talk about her husband, much less to him. She did not want to avoid experiences, thoughts, or conversations that might bring up the pain of her loss. Instead, she sought these out. On the first-year anniversary of her husband's death, she traveled to the remote vacation spot where they last saw each other as a way of reconciling what had happened. The letter was another way of confronting her pain and acknowledging the ongoing relationship that she would have with her husband.

In writing to her husband on anniversaries, birthdays, and other meaningful dates, Monique pushed the envelope of social media. She chose the most public and open way of communicating on Facebook; she could have created a small group group of sympathetic friends or written only on her husband's wall. Her public postings made some people uncomfortable, but this openness allowed both her close and extended friends to acknowledge her continuing emotional engagement with her husband.

In connecting with these friends—the ones who have responded with love and understanding instead of questions about her mental state—Monique has connected with the qualities of her husband that she sees in them. She finds continuity in their bonds with her husband and their memories of him.

Social media is creating new spaces and norms for mourning: compared to traditional forms of grieving in Western culture, mourning is now less confined to specific times, places, and people.[3] Mourning through social media may support continuity while pushing against norms of grieving oriented toward closure and "letting go."[4] Some scholars have voiced concern that mobile and social technology may either abbreviate or prolong grieving by offering emotional shortcuts to confronting death, preventing temporal compartmentalization, or forcing individuals into an ongoing performance of bereavement.[5] Comments on a deceased person's page by trolls often create additional pain for those in mourning, and this type of online harassment by total strangers added to Monique's suffering.[6] But Monique and others also use social media to create valuable continuity. Monique does this not by talking about her husband on social media but instead by addressing him directly. Their dialogue lives on.

Leaning into Grief

After her mother's sudden death, Dana was distraught. Her therapist suggested that Dana track her emotions as a way of "leaning into grief."[7] Dana tried various mood-tracking applications. These apps generally prompt one to pick a mood from a checklist or rate the intensity of particular moods, at whatever interval makes sense. Inspired by positive psychology, most of these apps are oriented toward increasing positive moods and decreasing negative ones—with the presumed motivation of getting happier.

But Dana's was not a happiness project. She wasn't trying to eradicate her grief. She wanted to delve into her mourning, to be with herself in relation to her mother. She also wanted to capture the emotions as *she* felt them. Tagging and scoring categories of moods would not help her articulate her feelings, thoughts, and perceptions.

So instead, she put together a combination of tools—Google Forms, an Excel spreadsheet, and the Flickr photo-sharing site—that let her record in her own words and images what she called her "sightings" of her mother: the occasions when memories of her mother came flooding back, and the times that she could most vividly feel her presence. She wrote notes on her phone about her experiences and their context, which were linked to the Google Form, and took pictures of the objects that reminded her of her mother. One was of a mere water bottle she saw at the gym. It was identical to the one that Dana had used every day when her mother was in the intensive care unit. At first the bottle was a trigger for grief, but as she uploaded the image and sat with it, it became, "like Proust's madeleine," she told me, a stimulus for reverie.

She also experimented with visualization software, developed by the QS pioneer Anne Wright, that allowed her to integrate and analyze the diverse forms of data that she had been collecting. She was open to spotting new patterns in how she grieved, but it became increasingly clear that this distracted from a simpler goal: remembering her mother.

Although her mother had been a mentor to many, Dana feared that she would be too soon forgotten. In fact, Dana worried that she herself would forget. Their relationship was defining for Dana. Her mother, a surgeon, was also her trusted mentor as she started her own medical training. Perhaps in her effort to be productive and plow through medical school, she would grieve too efficiently. So she tracked her mourning—not to avoid the pain, but to move toward it. She focused on the simple acts of capturing a moment, like the water bottle, and reflecting on it.

The act of tracking created an openness to the loss—a moment to commemorate her mother and their connection. These moments were at least as valuable to her as the analytics that she originally set out to obtain. She ended up with a memoir not of her mother but of her own mourning, a memoir of herself in relation to her memories of her mother. Tracking let her explore her loss, and reflect on her identity.

A year and a half after her mother's death, when we spoke, Dana was still open to grappling with questions of who she was becoming in relation to her mother. She had continued to move closer to her mom and her values, but she was also considering career directions that would move her further from her mother's path as a surgeon. Dana's tracking project surfaced questions rather than answers. But she was never seeking closure.

Throughout, Dana measured herself in a way that was personally meaningful, defying trends in self-tracking practices and mood applications. In addition to broadening her effort from tracking herself to tracking herself in relation to her mother, Dana challenged the notion that the value of tracking comes from discovering patterns in longitudinal data and pursuing goals. For Dana, the value was more in the immediate act of reflection than in longitudinal analytics. And whereas most mood-tracking apps focus on lifting mood, Dana embraced her despair. She tracked in the spirit of expanding her emotional range, including the deep pain of loss.

Shoebox Memories

In her last year of college, Amanda fell into a relationship that moved at a dizzying speed. Within two months of meeting, she and her boyfriend were talking about eloping. He was a couple of years older and about to begin a graduate program in a different state. Focused on finishing her coursework and leading a student organization, Amanda didn't mind that he was making many of the decisions in their relationship. She didn't have plans for what she would do after graduating, so their relationship became her plan.

When Amanda realized that the relationship was getting so serious, she wanted to memorialize its start. She scrolled up to the beginning of their voluminous SMS dialogues and took a screenshot of their first exchange. She grabbed other pivotal moments, such as the first time he used a heart emoji. From that point onward, she captured the minutiae and milestones of their relationship through photos and screenshots of texts, Snapchats, and anything else that occurred to her.

Then the relationship ended abruptly, while Amanda was thinking of little else. For months afterward, she scrolled through these photos, reminiscing about their relationship. Sometimes she was searching for the photos, and sometimes she stumbled on them as she was trying to find something else. The images evoked painful memories of their tumultuous breakup. At one point, when she was running out of storage space on her phone, she tucked some of the images into a folder on her PC. Although not explicitly her intent, moving the images prevented her from looking at them.

She continued to be haunted by her ex's presence on social media, however, even long afterward. Amanda recalls a trip she took with the friend who introduced them. She and her friend posted pictures that were almost identical. They were both in all of them. Why, she wondered, had her ex only "liked" her friend's photos and not hers? She realized this puzzling was an unhealthy indulgence, but she struggled to redirect her attention.

One night, things shifted. She and several friends were glamming up for a party. As they were picking selfies to share on Instagram, Amanda caught herself thinking about which ones her ex would find attractive. Even as she was enjoying her friends, admittedly looking fantastic, and headed out to meet new people, she was still worrying about his impressions of her. She needed this to change. On that night, over a year after their split, she blocked him from viewing and commenting on her posts.

Looking back, she has mixed feelings about how she managed all of this, particularly the screenshots of their early online social interactions. She sees that she tortured herself unnecessarily by looking at the images so often and wishes she had moved them out of sight immediately. But she pointed out that when she was at her lowest, it was difficult to make active choices about what would be helpful, much less to follow through on them. "Did I

think it would help me to look at the pictures? No. Did I think it was helpful to lie in bed? Of course not." At that phase, though, it was hard to take
care of herself; her instincts were to move into the pain.

For Amanda, there is a strong tension between memorialization and forgetting. She thinks a lot about what she might want to look back on later
and thus resists throwing away images that capture important moments.
She has moved on but doesn't want to erase the relationship from her history. This urge to memorialize was heightened on November 11, 2016, when
Facebook erroneously categorized her, Mark Zuckerberg, and many others as
as having died.[8] Her page was overwritten with boilerplate text about fond
memories from mourning friends and family. Although she laughed at this,
the mention of her mortality made her even more aware of key moments
that she or others may want to look back on.

It also made her a little more intent on deleting pictures that embarrass
her now—ones that might embarrass her in the future and not represent
her well after her death. She cares about what she leaves behind and who
will see it. She cares about who will get her phone, too. It is rich with data
and could offer someone a real sense of her experience. She doesn't want this
going to waste. Recently, her current boyfriend added her fingerprint to his
phone and asked that she reciprocate. She declined. But she thinks about
giving this access to a good friend who might help her get through a difficult time or could really appreciate her phone's contents when she is gone.

"Instead of a box of love letters under my bed, I have a collection of
screenshots of text messages from former flings in the photo album of my
iPhone," says Amanda. The upside of a physical shoebox is that love letters
are hidden from sight in daily life. Phones, on the other hand, keep these
traces at our fingertips and in our sight. These digital devices make it easier
to remember and harder to forget.

And unlike shoeboxes, these devices encode what they have seen. The
screenshot captures the phone's experience as well as our own. What might
have been ephemeral is archived and annotated with contextual cues such
as the time, battery life, and signal. Casual text messages that no one knew
were the beginning of a deep relationship, along with social media posts
and even something as mundane as directions, are often captured as photos,
and these survive the entropy that would otherwise ensue.

A student in my class who read Amanda's scenario described a strategy
used among her friends. When one of them suffers a breakup, her friends

seize her phone and lock all the content related to her ex in a password-protected folder. After a year, they will agree to share the password with the sufferer. She believed this intervention had been successful since, at least in one case, a friend forgot to ask for the password after a year. She had no qualms about automating this kind of help, envisioning a "delete a friend-ship" option on Facebook. In fact, she would like a more systematic erasure than the blocking options currently available, or even than the Pandora's box envisioned by design researchers Corina Sas and Steve Whittaker to hide digital possessions from view until an individual has recovered from the breakup.[9] Instead, this student imagines a single click option to purge all media, texts, email, and other relationship traces. She doesn't want to see them again.

Where once forgetting was inevitable, our devices sometimes now require us to take active steps to stop remembering. Should we hand forgetting back to our devices by installing systems to prevent torturous reverie? It is easy to imagine algorithms that identify emotional pain via cues in our words and images, or other posting behavior.[10] Having identified it, the software could then compartmentalize the painful content, hiding it from our view. Fictional depictions of these ideas, for example in the films *Rememories* and *Eternal Sunshine of the Spotted Mind*, run alongside medical science on treating traumatic stress.[11] We will have choices about how we use technology to forget, whether that is through some kind of automatic protection from painful memories or for active rituals of separation.[12] We should consider if we really want to leave emotional recovery from rejection and other pain in the hands of the Facebook's "Compassion Team" or if enlisting the assistance of friends, perhaps with established policies in advance, is preferable. And perhaps there will be ways to combine strategies, giving our friends permission to enable AI on our behalf.

Remembering the Context

While some struggle to forget, others struggle to remember. When I met her, Claire was in her early seventies. She moved with some effort and bewilderment, as if she were showing me around someone else's home rather than the one that she had lived in for almost fifty years. Occasionally, though, I caught glimpses of the tall, strong woman who previously held command over the house. And I got a sense of her strength in the way that she discussed her concerns about her failing memory. She was pained by her forgetfulness, aware that she increasingly struggled to do multiple things in parallel or accomplish tasks involving multiple steps. It was getting harder for her to host family events that she used to pull off with ease. She worried that this was the beginning of a truly debilitating loss of memory.

Claire proudly showed me one of her prompting techniques: a large, bright yellow timer she wore around her neck. She turned the timer on when she started the laundry so that she would remember to put it in the dryer. In the intervening thirty minutes, she walked me around her home, pointing out the organization of her kitchen along with the lists she kept for meal preparation and shopping.

When I looked at her to-do list, the timing seemed drawn out. Tasks that she used to manage in parallel the day of a family dinner were now spread out over the preceding week. Asparagus was being prepared on Wednesday for a Sunday dinner. No one else acknowledged the decline in her cognitive health—not her husband or kids, and not her clinicians, who would have needed to see signs of impairment on cognitive tests along with self-reported changes. Claire performed satisfactorily on these tests, but they didn't offer a comparison to her own baseline, just to population norms. The denial of her husband and children made it hard for Claire to draw them into her attempt to get a diagnosis or as helpers in creating memory aids.[13]

After thirty minutes, the deafeningly loud buzzer around Claire's neck went off. She looked at me and asked, "Now what could that be?" She picked up an unopened mobile phone that she had received as a gift and wondered aloud if it could be creating the sound.

Memory is highly contextualized. The buzzer around her neck was not intuitively attached to the dryer. Contrast this with Post-it Notes stuck to doors, drawers, and medicine cabinets by a caring daughter or son to remind an elderly parent to put on a jacket before exiting, eat something in particular for lunch, or take medicine—a tactic I've seen in almost all my home interviews with cognitively impaired adults and their families. These notes are thoughtfully placed in the physical context of the tasks they prompt. They are not a perfect solution; an elder isn't always near the medicine cabinet when he or she is due to take a pill. But when we develop prompts, it helps to embed them in associated objects, contexts, and relationships.

In the years since this interview, technologies have become increasingly connected. We can now associate our coffee makers with alarm clocks and front door with the thermostat. Some dryers are already capable of sending texts when they are finished, and we can tag our keys and other objects that are prone to wander. Some people are using tools like Amazon's Alexa to set reminders, tools that will become easier to customize and integrate with other devices, such as doorknobs or stoves. Whether we are sticking up Post-its or developing scripts for "smart things," the important thing is to be specific in our prompts and, when possible, involve others in making them meaningful. We should tone down unnecessary alarm; a generic loud noise, say, may just create anxiety. When the house is not on fire, a calm statement that "the laundry is done" or signal that intuitively maps to that task will suffice. Our sensations search for associations. A reminder needs context to have meaning.

Calm Alarms

Anyone who has nervously searched for their keys, or the eyeglasses that were sitting on top of their heads all along, knows the panic that comes with forgetting. The reminders we create for ourselves shouldn't add to this anxiety. They should be calming. This is true whether one is struggling with cognitive decline, as Claire was, or just managing a busy life. This was the case for Lance.

Lance and Jacqueline, who have been living together for fifteen years, have few household disagreements. They are scientists who value both order and creativity. Jacqueline appreciates systematic organization and planning, whereas Lance plunges into new projects without caution. Their home feels like a busy and well-organized lab. There's bustle, but nothing feels out of control. Lance is frequently experimenting with new gadgets for research projects and daring futuristic recipes. Storms of activity in the kitchen or office are followed by efficient and thorough cleanup. Many packages arrive with ingredients for these endeavors. After the contents are emptied, the boxes are immediately folded and put in the recycling bin.

Lance is responsible for bringing out the recycling every other Tuesday night. He always intended to follow through with this task, but for a long time the irregular and infrequent schedule made it easy for him to forget. The same conversation played out every week, verbatim, becoming slightly more annoying to Jaqueline each time. Realizing that it was a Tuesday, Lance would ask Jacqueline, "Is it recycling or trash this week?" This minor, recurring interruption frustrated Jacqueline, in part because she wanted him to take complete responsibility for the task.

To limit the friction, Lance tried a visual reminder: using IFTTT, he linked a Google Calendar he created for this single task to the programmable lights near the trash and recycling area. The lights now turn green on recycling nights. On the other Tuesday evenings, when only trash and compost are taken out, the lights turn blue. When he goes to take out the garbage on Tuesdays, he sees the light and knows whether he needs to take out the recycling as well. This contextual reminder is just prominent enough to ensure the correct task is completed; he is not interrupted from other things by a jolting buzz or clang.

Jacqueline no longer needs to worry about whether Lance will remember to take out the recycling or feel resentment that he did not. He no longer feels that Jacqueline questions his conscientiousness. They can both appreciate the lab and kitchen wizardry without worrying about the aftermath. The ambient notification, meaningfully placed, effectively preempts tense exchanges. Their system signals just enough: it doesn't shame or shock; it calmly reminds and reinforces trust.

When Remembering Is Not Enough

Reminders can be helpful, but what recourse do we have when they're not enough? For people who struggle to change ingrained ways of eating, habits and sometimes cravings can override even the most memorable intentions. Mary is in her sixties with type 2 diabetes. She lives in a small house in Los Angeles with her husband, also diabetic, her granddaughter, and great-granddaughter.

Her realization that she had diabetes came in two phases. First was what she described as "five years of denial" after she received the diagnosis during a routine exam. During that time, all she did was switch from Coke to Diet Coke. She did not give up on her everyday high-glycemic fare, much less items like cake that had always been a part of family dinners. Then one day at work, she began sweating profusely and was brought to the emergency room. Her blood glucose had shot up to six hundred, far beyond the upper boundary of normal. "There was no more denying that I had diabetes," Mary told me, explaining how at that moment she resolved to make drastic changes to her diet. She would eat in a balanced way, monitor her blood sugar, and modulate it through diet and medication. "There was no more denial," she emphasized. I was convinced.

But then Mary received a call from her husband, who was at the grocery store. I was sitting a couple feet away, with a videographer behind me. Her husband asked what he should pick up, tempting her with options from the entire grocery store, not just those low on the glycemic index. "Black-eyed peas, potatoes," Mary listed, and then after a slight pause, "and German chocolate cake."

When her husband called, presenting this immediate gratification, Mary's attention shifted from health to indulgence and family traditions. This is not because Mary is weak. Behavioral economics researchers have demonstrated that we consistently opt for short-term rewards, even if they are measly compared to ones that pay off more significantly, over longer periods of time. This disconnect, called temporal discounting, shows how our values change depending on time and situation, and how difficult it is to pursue abstract rewards (such as health) when up close to alternative rewards that are concrete (such as cake).[14] Research has also shown that knowing our vulnerability for surrendering long-term interests to impulses doesn't protect us from doing so. Insight is not enough.

To align behaviors with long-term goals, some take seemingly extreme actions, such as betting with friends that they will stick to a diet, painting fingernails with bitter-tasting polish to deter biting them, or disabling wireless on a laptop to focus on work.[15] One of the more effective strategies is to set the desired farsighted actions as defaults, such as opt-out retirement savings plans.[16] Policies tailored to an individual's long-term goals, for example, eliminating all prepackaged foods or instituting a forty-eight hour waiting period before any significant purchase, can be helpful too.

Sadly, many efforts at self-control can backfire. Rigid policies can impair the attentional and emotional resources necessary to pursue rewards. As psychiatrist George Ainslie states, "If you have yourself tied to a mast, you can't row; if you block attention or memory, you may miss vital information; and if you nip emotion in the bud, you'll become emotionally cold."[17]

The optimal approach involves flexibility. Psychologist Howard Rachlin advises resisting our vulnerabilities with "soft commitments" rather than rigid policies: building momentum out of positive habits and viewing our choices as patterns rather than as a series of independent decisions.[18] Take the choice between reading several pages of a book before falling asleep or a dozen status updates. The choices may seem similar in that moment. But if you compare the experience, say at the end of the week, of finishing the book to having perused seventy-plus disconnected status updates, the picture changes; in the long term, the wholeness of the book most likely adds up to more. Further, simply starting the book makes it more likely that it will be chosen the following evening; our actions create momentum.

Understanding that any choice is predictive of future choices helps us balance the consequences. A slice of cake today makes it more likely that we'll have one tomorrow. Reckoning with these rules of behavior could help Mary resist asking for chocolate cake and dismantle the notion of indulging in "just one last slice." In general, we can start to see how some of the things that bring about immediate pleasure can erode our long-term happiness.

Mary's technology, especially her use of the phone, did not help her stay committed to her future health. Her husband's call from the store removed some of the natural barriers to temptation, such as the effort to go shopping herself. The last-minute request bypassed deliberation and planning. The phone increased her vulnerability rather than supporting her long-term objectives. Although Mary did not adapt the phone or any other technology to align daily shopping choices with long-term health goals, her experience raises the question of how one could.

One way to work against temporal discounting (the tendency to devalue long-term incentives when we approach a short-term reward) is to set policies in advance.[19] For example, Mary could set a recurring shopping list—a default—and ask her husband not to call her from the store. The list could be made on the back of an envelope, in an email, or using voice assistants like Amazon's Alexa or Google Home, which will presumably get better at generating lists based on what's running low, what's in season or priced well, who we are eating with, and what aligns with our health goals. These degrees of automation would remove some opportunities for self-sabotage. It would be harder for Mary to add items that are delicious but dangerous. Whether it's created automatically or manually, a list can be shared and vetted by a knowledgeable peer, coach, or other health care professional. Third-party approval of a list, explored further in chapter 8, might help a couple like Mary and her husband engage as allies in managing their health.

One Space, Two Places

Like Mary, Ellen didn't have trouble remembering exactly. She had trouble changing her behavior to align with a goal that haunted her: she had repeatedly resolved to stop eating at the end of meals and cease snacking between them. Ellen had never been overweight, but she had an obsessive preoccupation with eating and had worked for years to manage it. Setting intentions hadn't helped her. What did help was changing her environment. She kept her kitchen empty, except for some tea in a cupboard. She knew that if she kept food at home, she would snack incessantly. She had been through cycles of extreme dietary restrictions and excessive violations. Many years earlier, a friend with similar struggles suggested an approach that had worked for her: eat all three meals out, ideally with others. That worked for Ellen, too.

But then her relationship with David got serious, and they moved in together. David wanted to be supportive but was frustrated. She knew what he must have been thinking: How could someone lack the control to stop eating, especially something as bland as crackers? He had previously lost over fifty pounds without emptying out his kitchen or inconveniencing others. He wanted the option of eating breakfast at home, having snacks he could grab from the cupboard, or preparing a meal without going to the store. Ellen didn't want to withhold these basic pleasures from him and was ashamed of her lack of restraint. The stocked kitchen was wreaking havoc on her eating, though, unraveling years of progress.

At the same time that Ellen and David moved in together, she started working from home. Now there were many more hours of the day that she had to contend with the kitchen. And there was always some kind of stress—ambiguous assignments or contentious emails that she would ordinarily bounce off the colleagues sitting beside her. Almost everything that showed up on her computer added to her stress. In the cupboards, she sought some kind of buffer zone where she could process things before responding. But the grazing was becoming as much a cause of anxiety as a response to it. She felt terrible about her lack of control.

She asked David to keep certain snacks in his bag or car instead of the kitchen—a mobile mini-mart to which she would not have access. When he rejected that idea, she wondered aloud about a lock for one of the cabinets. David shuddered to think about the effect that this would have on their own interactions, let alone the awkwardness that would arise with guests.

David had been experimenting with magnetic locks in other projects and began thinking about how they could work with the do-it-yourself "maker" programming tools that he often used. After researching electro-mechanical locks online, he found one that had been used for cattle fences. Both components of this lock would easily fit inside the cabinet. He then programmed a Raspberry Pi (a tiny computer commonly used in the maker community) to control a relay that switches the magnetic lock on and off. All this is tucked away in the corner of the cabinet. The lock is operable from an app on his phone, which allows him to control it not just from within the house but also remotely. He can use the phone discretely as a key, and he can also give access to others on their phones. The kitchen is now safe for them both.

The system is not a cure and may not stay in place forever, but it has made a significant difference for Ellen and David's relationship. When David is home, the cabinet is open. When he is at work, it is locked. And when it is locked, Ellen does not obsess about its contents. She can focus on her work, without the escape strategy of snacking. The lock is not visible to them or to visitors. Because it is subtle and not visible on the outside, they don't discuss it much.

One of my students who read this scenario strongly objected to Ellen's behavior and needs. He thought they were investing in "a Band-Aid rather than a cure." But a Band-Aid intervention like this may play an important role in healing for a couple of reasons. To begin with, it permits Ellen the intimacy of living with someone else. Previously, she dreaded the idea of trying to reconcile her eating habits with someone else's. Eating disturbances, like most psychological struggles, are thought about in terms of interpersonal dynamics as well as neurobiology and other psychological factors. Loneliness may contribute to eating disorders, and the secretive nature of the disorders may heighten feelings of isolation.[20] Interventions often involve decreasing shame and isolation through group, family, and, most pertinent to this story, couples therapy.[21] In Ellen and David's case, improvising this temporary strategy and talking about eating concerns in practical terms made Ellen feel like less of an oddity.

And the solution may ultimately run deeper than a Band-Aid. It is conceivable, for instance, that the flexibility and security experienced from

living with someone else may generalize to Ellen's eating habits. By helping
to create this solution and using it, she expands her repertoire of coping
strategies. In shifting from an absolutist policy of avoiding any exposure
to food in her home to a more modulated control, she tolerates more risk
and ambiguity.

Ellen and David's story sits atop significant data about how our environments shape our health decisions.[22] For example, studies have shown that
people from countries with less fast food gain weight after moving to the
United States; people walk more in dense cities than in suburbs; and avoiding cues associated with substance abuse helps those in recovery from addiction.[23] Changing the environment or system surrounding an individual is
often a viable way of improving health, with evidence found for exercise
classes in school, cigarette taxes, public smoking bans, and urban design to
increase exercise.[24] To deter impulse purchases of unhealthy snacks, some
researchers have even explored the effects of imposing time delays in vend-
ing machines.[25]

Ellen similarly changes her environment rather than relying on deter-
mination or intention to change her behavior. There's some wisdom in this
approach given that intention gives way to habit, and habits can most easily
be shifted through changes in context.[26] She locks herself out with simple
technology put in the hands of a trusted partner. Her strategy resembles
that of gamblers who request that casinos place their name on a blacklist, or
Ulysses tying himself to a ship's mast so that he would not act on the temptation of the Siren's songs.[27] And although this may seem like a prototypically
rigid policy, which was warned against in the discussion above, it is more
flexible than Ellen's previous strategies.

Ellen and David's challenge arose because what we think of as a single
space is actually a different place for each person who inhabits it. The stocked
cupboard that calms David creates stress for Ellen. Similar conflicts play out in
other households. One woman with diabetes I interviewed finds it difficult
to limit carbohydrates because she shares her home with her mother, who
has Crohn's disease and subsides on bland cereals. Another diabetic, Mary
from the earlier in the chapter, faces temptations not just because her husband calls her from the supermarket but also because her granddaughter, who
lives with her, insists on sugary cereals. A vegan teenager may be repulsed
by the smell of the pot roast cooking for the rest of the family, and so on.
One space can be radically different places. Connected devices may be able

to help household members negotiate different needs and perhaps carve out places within the same space.

Of course, that relief can come only if the people dwelling together acknowledge differences and share an interest in supporting one another.

Imperfect Recall

I use the terms remembering and forgetting loosely in this chapter to include the many ways we focus our care and attention. Some of these stories are about remembering important people and continuing conversations with them after their death. Some are about remembering the conversations we've had with ourselves and holding ourselves accountable to the promises we've made. Two of the stories explore the practical cues for remembering the tasks we are prone to forget. In other stories, individuals struggle to forget—an ex in one case, and tempting snacks in another. Many of the individuals refocused their attention by changing their physical or digital environments. This invites us to think about how we shape the environments that will end up influencing our own actions, often in ways that we are not even aware of at the moment of a choice. It is worth trying to design these spaces so that they guide us toward what we want for ourselves in the long run.

5 Beyond the Hookup

Tinder is often presented in the media as a wildly popular tool for sexual encounters with no strings attached. Its matching—based solely on mutual physical attraction and location—almost mocks the compatibility algorithms of legacy online dating tools, such as Match.com and eHarmony. Headlines such as "How Tinder Took Me from Serial Monogamy to Casual Sex" were common, particularly in its early years.[1]

The actual use of Tinder and its gay equivalent, Grindr, is far more textured than this portrayal. Since Tinder's release on college campuses, it has been taken up not just for convenient casual encounters but also for establishing ongoing relationships and meeting friends.

For the people you will meet in this chapter, hooking up was beside the point. They used Tinder and other hookup apps (sometimes in conjunction with social media) to explore their sense of self and place in various social worlds. Even the cases where apps helped them locate a romantic partner, their motivations were never that simple.

Bouncing Back from a Breakup

Caroline, a twenty-two-year-old female college student, felt broken when her boyfriend of four years broke up with her. This came within months of emotionally significant milestones, such as introducing him to her father, whom she rarely sees, and exposing him to conflicts within her family. She thought that this sharing reflected the strength of their relationship but afterward saw that it did not carry the same meaning for him. For months after the breakup, she struggled. Her friends sometimes found her crying in the back staircase of her sorority house.

Unlike her family troubles, which she kept hidden from most of her friends, this breakup pain was something she shared. A friend, who had witnessed the suffering and listened to Caroline talk about her pain at length, was unsure how to help. More empathetic discussions didn't seem like the answer. So during one of these laments, she grabbed Caroline's phone and downloaded Tinder. She browsed Tinder frequently even though she was in a committed relationship and was optimistic about how it might help Caroline.

That wasn't the commiseration that Caroline was expecting, but it worked. Excitement overtook her despair as she browsed matches. She described the charge: "When in real life would I get ten messages saying, 'That guy who you thought was cute, well he thinks you're cute too'?!!" She used the app as a form of social buffering. It ameliorated the pain of being dumped and created an opening for excitement.

She used the app in some slightly uncustomary ways. Tinder promotes its link with Facebook, in part to provide assurance about the identity of other people on the app and in part to pair up with people within their social networks. An identity is less likely to be fabricated on Tinder than on other dating sites (although some do create alternative Facebook accounts to disguise themselves on Tinder). To some, hooking up with mutual friends seems appealing and less dangerous than meeting strangers, but not to Caroline. She avoided any matches with mutual friends. Most of all, she didn't want her sorority sisters involved in this aspect of her healing.

While many people take advantage of Tinder's geolocation features to find potential matches nearby, Caroline set wide location parameters purposefully to avoid meeting anyone from campus or university circles. She focused on low-income suburbs a good distance from school. Doing so was also a way to visibly reject her family and the high value that they, especially her mother, put on wealth as a criterion for selecting a husband. Tinder became a weapon in Caroline's continuing struggle with her family as well as a way to heal emotionally—psychological uses that may not immediately come to mind when one thinks of Tinder.

Beholding the In-Between

Paula, a twenty-five-year-old grad student who moved to the United States from the United Arab Emirates, has used online dating apps in a deliberate way to understand and navigate cultural identities. On paper, she is an

Indian citizen, but she thinks of herself as a "third-culture kid," trapped between multiple cultures while never truly belonging to any of them. Her experience is not that of first-generation Indians like her parents, who adhere to traditional Indian culture, nor is it that of second-generation immigrants, who can often legally and culturally identify a country they live in as their "home." For as long as she can remember, she has been hyperaware of the multicultural dynamics in her environment. Her parents moved to the UAE from India for her father's work, and she lived there through high school, attending an international school with mostly British teachers and students from across the world. The complex social hierarchy, with obvious bias against Indians, contributed to an internal struggle for Paula. She felt conflicted between reluctantly accepting her Indian heritage and conforming to her friends' Eurocentric culture. When she thought about relationships and the kind of marriage she would ultimately want, she considered the dynamics of compatibility and culture, and how she could avoid a marriage of conflict like that of her parents. For even though her parents had an unconventional interfaith "love marriage," it struck Paula as far more conflict ridden than the arranged marriages of her friends' parents.

With all this at play, Paula started out solely dating non-Indian men. In college she had a Latino boyfriend but broke this off as she moved away for graduate school. The relationship would have ended even if she had stayed. He expressed no interest in learning about her background and made racist comments, sometimes mocking the speech of his Indian colleagues as if she would find this amusing. Days after their breakup, she got on Tinder. She wasn't looking for a hookup; she feels attraction when she sees someone as a prospect for a long-term relationship. For her, Tinder was a place to understand the city that she had just moved to and her place in it. It was obvious from the start that because her criteria had become fairly strict, her options would be limited. Partly through bonding with friends of similar backgrounds in college, her goals had become clear: to work in the United States, she needed to navigate the complex immigration system; getting married to a US citizen almost felt like a necessity. That person also needed to be attractive, well educated, and professionally ambitious.

After spending some time on Tinder and other apps, her criteria sharpened further. She became less tolerant of ethnic bias and quicker to spot it. Paula flinched when she read comments based on her picture such as "OMG, you are so exotic." These remarks highlighted the ignorance of men who didn't understand anything about her background. It was even more appalling to her when they came from men who had traveled extensively. She noted that she never received these comments from men of Indian descent. It was becoming clear that the Indian men she met online tended to meet her other criteria as well. It wasn't her plan, but shared ethnicity was becoming a proxy for what she wanted.

So she started dating Indian men almost exclusively. Other compatibility challenges arose quickly with this approach, however. She broke off one relationship with a man after realizing just how Indian he was, even though he was a second-generation immigrant. This became clear when he told her that he would probably never tell his Hindu parents about their relationship because they would object to her interfaith background. His parents' conservatism, for example, their acceptance of the caste system, clashed with her beliefs. His accommodation of their values, in his refusal to tell them about the relationship, suggested rifts to come.

After their breakup, she spent time online, carefully screening profile pictures for indicators of socioeconomic and educational compatibility. She dated a couple of people during this time, but nothing stuck. She and her current circle of friends—all of whom have lived in multiple countries, speak many languages, and are "some degree of atheist"—have the same criteria for men and regularly discuss their searches. They immediately ban anyone whose profile indicates a vastly different cultural milieu from their own, sometimes conveyed through cues such as a mirror selfie. They want to date Indian men but steer clear of first-generation immigrants, who they suspect will be "too Indian," meaning highly religious or culturally conservative. She can usually discern acculturation from profile pictures; however, she also scrutinizes writing for signs of nonnative fluency—overly abbreviated messages and specific errors such as a space before an exclamation point— that suggest incompatibility.

She met her current boyfriend on Tinder, inferring from his picture that he met most of her criteria. She seemed to assess his profile accurately. Even though his messaging on the app was awkward, she liked him in person. Perhaps most important, he shares her third-culture experience of not fitting particularly well into either Western or Indian culture. They have been together for a year, and, although her assessment of compatibility is ongoing, she's optimistic. She can talk with him about the nuances of her history, such as her parents' languages and faiths, and the evolution of her own beliefs, all of which he can relate to.

So an app known for the quick hookup is, in Paula's hands, a way of reading complex signals about ethnicity, attitudes, and acculturation. This lets her bring the complexity of her own in-between cultural experiences into the relationship, validating their significance to her.

Grindr Is the New Airbnb

Like Paula, Thomas used Tinder to explore his identity. The previous year on a study abroad program, Thomas started dating men. Back in his hometown for summer break, he knew only his straight high school friends. He longed for the acceptance and romantic possibility that he felt while traveling. Tinder and Grindr helped.

He did not meet his Tinder matches in person. But the evidence of desirability—the fact that men he found attractive also found him attractive—was validating. He found meaning in what he called "people closeness," Tinder's indication of mutual friends. He had the sense that he could have met these matches through a casual introduction in the course of daily life. This social proximity made their interest in him more salient but also more daunting. He thought about these friends of friends not as hookups but as prospects for more serious relationships, something that he was not yet ready to pursue.

In this period of exploring his identity, Tinder and to a greater extent Grindr allowed him to express his sexuality—something that he had silenced until then—and think of himself as a sexual person. Grindr's menu options radically simplified a complex social world. It was pretty easy for him to select the boxes that fit his perception of himself and what he was seeking, and to experiment with different configurations of selections and search results. Grindr and its checklists opened his eyes to the good and bad. Its pervasive racial and body-type biases disturbed him, but the app also allowed him to map out this new world and his place in it. During a period between knowing that he was gay and becoming actively sexual, it was a place to experiment with presenting himself in a way that gave him evidence of his sexuality and attractiveness.

Down the road, Grindr became a pivotal tool not just for hookups but for meeting friends. When we last spoke, Thomas was in Mexico City, enjoying clubs and restaurants with people he had met through the app. The culture of a place of course affects how flexibly an app like Grindr is used. While in Abu Dhabi, where homosexuality is a crime, Thomas noted that the app was confined to discrete hookups, but in Hong Kong it was understood to be a general-purpose social navigation aid that gay men used to meet up for many reasons. His time in Hong Kong was completely shaped by the friends that he met through Grindr. Reflecting on his overall experience with the app, he described a sense of possibility that extends far beyond the encounters that Grindr is most known for: "I can find a relationship anywhere in the world."

Winning Tinder

Archie proudly announced on Tinder that he had a thousand matches. "That means I won Tinder, right?" he joked.

If "winning Tinder" meant anything, one might expect that to be an exciting series of sexual or romantic encounters. Those in-person encounters are what the wildly popular app was designed for. Many online daters have gamed the system to obtain better matches, and some have shared their success strategies in the form of plug-ins, secondary apps, and books.[2] These successes are generally about optimizing one's profile to increase matches with people the individual considers attractive and ultimately to have more face-to-face encounters with those matches.

But when Archie, a man in his twenties, posted a screenshot recording the number of matches that he had accrued, he was after something different from hookups. He had not met any of his matches and did not intend to.

When asked why he uses Tinder by a researcher on our project, Archie explained, "Probably some combination of trying to entertain myself, trying to bolster my self-esteem, and just enjoying scrolling through pictures of attractive girls for fun." He said that his self-esteem has always rested on people wanting to talk to him, as he has always been hesitant to make the first move. "So like when a girl messages me first on here, that's probably the biggest esteem boost."

He continued, "To be honest, I still find the idea of actually meeting Tinder people in person so weird!" His preoccupation with match count,

although playful, belies an attempt to measure up. The boast conveys his eagerness to accrue points—something that is harder for men than women on sites such as Tinder. It is similar to accruing Instagram likes or—for some people—airline miles, but more personal. The matches are not about his talents or clever comments but instead his personal attractiveness, or possibly a cute puppy, if he's included one in his profile photo.

He was using Tinder for something that it wasn't directly intended for—a solitary self-esteem game—adapting one of its features to do so: he posted the screenshot through Tinder's Moments feature, which invited users to share photos with all their matches for twenty-four hours. The photos intended for this sharing were probably not screenshots of one's own Tinder statistics.

Archie's boast of a thousand matches was intended to signal his playful "gamifying" of Tinder and presumably his desirability. He did view the app as a game, so in some ways this was an honest signal. And in some cases, a general reputation as attractive or likable can draw favorable impressions. He was, I think, trying to create a perpetual loop of validation. But any woman who reads this story can predict the outcome: some of his matches wrote to announce their irritation, and more than a few unmatched him. He had signaled that his matches were a collection, not individuals.

A strategy like this one for boosting self-esteem—chasing Tinder matches to see how many you can accrue—may work to some extent. It confirmed Archie's sense that he was attractive. But publicizing this game would obviously offend people who see themselves as more than points. His strategy was fragile for other reasons, too. His matches were based entirely on quick ratings of photos. He and a thousand women swiped right to indicate mutual attraction. What if Archie posts a photo that no one happens to like? What if over time people find him less attractive? Archie may have understood that the matches were not firm ground on which to base his sense of self-worth, and that validation based on enduring characteristics or efforts would be more stable.[3] Perhaps that was why he categorized the pursuit of a thousand matches as a game.

The Tinder Queen

While Archie boasted about winning Tinder, Camilla proclaimed herself a "Tinder Queen." She hadn't always felt respected on dates she met offline, but on Tinder she feels in control. She works at creating a glamorous persona and regularly curates the Facebook photos and interests that show up on her Tinder profile. She wants to meet people, or at least accrue matches, wherever she is, so when she travels, she modifies her profile to express what she thinks will be appealing in that context. For example, she shows more playful images when on spring break than when she's interning (e.g., sporting a tank top and sunglasses at an outdoor bar as opposed to being suited up in an office). She treats her profile picture as if it were a status update, adapting it to her goals for a particular situation. She noted, confidently, that she tailors her messages to the people who write her. She uses language from their messages and profiles, understanding that this kind of mirroring can make one more likable. This could backfire, though; mirroring is effective only if it is not obvious to the recipient, and some of her mirroring, such as throwing in expressions from the other person's native language, is likely to be noticed.[4]

Some viewers might be struck by the disjointedness of Camilla's self-presentation over time. A match from her home city might be put off by how she has changed her profile during a trip. She has had these kinds of complaints, but they've seemed laughably provincial to her. Creating a continuous persona is less important to her than cultivating new matches. Like Archie with one thousand matches, she sought evidence of her desirability. As my colleague and I sat with her, she mocked some of the overly earnest responses that she's received, and, swiping through images, commented, "These aren't real people." It seemed as if Tinder was a fantasy game where matches were points and confidence was the goal. She recounted a time when she sat around playing on Tinder with a couple of her friends. "It was like we were the three most beautiful women in the world."

Camilla demonstrates a fair amount of self-awareness and social skill. She recognizes that her own goals and those of others vary depending on context. She tailors her profile to what she wants at a given moment, and mirrors tone and language to gain acceptance—a practice that has been shown to build rapport in many situations, from dating to salary negotiations.[5]

Camilla's strategy is complicated. She exerts effort and skill to elicit the interest of people who she, for the most part, doesn't find interesting. Some may see this as a defense against disappointment or suggest that she focus more on quality rather than quantity of matches. But the evidence of her broad appeal, wherever she is at that moment, is clearly important to her. This evidence may be all she is seeking from Tinder.

Turning the Tables

The aggressive, escalating insult in the screenshot below is unfortunately typical. I found it not on a dating site such as OkCupid or Tinder but instead on Instagram, submitted to the account Bye Felipe. Like many women who have tried online dating, Alexandra Tweten had been harassed after declining or ignoring overtures from men. She started Bye Felipe to publicize this widespread problem, help women who have been harassed see that they are not alone, and help men grasp what women experience online.[6] Sexual harassment in online dating has been documented, but Tweten presents viewers with examples rather than statistics.[7] Initial amusement transforms into a deeper concern as one confronts the extent of these messages

and angry, often-violent threats within them. Bye Felipe's large following (approximately 430,000) speaks to the interest in bold interventions to address harassment.

The women who share such screenshots shift the shame onto their harassers. The screenshot is a particularly compelling means of outing the offender here, in part because the harassment occurred in an intimate channel with trappings of privacy. The offender had taken advantage of that one-to-one channel for statements that he would probably not make in the presence of others. The screenshot makes private harassment public. When one has read an actual exchange, it is harder to dismiss complaints of harassment as imagined, overblown, or somehow invited.

Often the user ID of the perpetrator is displayed in these postings—a revelation that could change that person's reputation. In these identity displays, Bye Felipe manifests a tension between protecting the safety of

those who are harassed and the privacy of perpetrators. A similar function—
a kind of virtual graffiti on women's bathroom stalls—has been served by
other apps, such as one called Lulu, which allowed women to exchange
insider information about men they had dated within their social network.
Lulu was shut down after three years, but Bye Felipe continues and has
inspired other forums for turning the tables on harassment.

Seeking Men with Faces

Curtis similarly created a forum for commiserating about his online dating disappointments. Curtis is a realist; he doesn't necessarily think that
he's going to find an exclusive, long-term relationship on Grindr. But he
wants his sexual partnerships to feel real with some measure of intimacy,
not anonymous or disconnected from the rest of his life. He doesn't exempt
the partners he meets online from the standards for kindness, humor, and
wit to which he holds friends, or the lovers he meets off-line.

On Grindr, he finds a style of communication that wouldn't pass muster
elsewhere. Most of the men who write to him include photos, just of specific
body parts. Curtis wants to see the person, to have a sense of mutual respect,
even if there will be only a single encounter. "I'm old-fashioned. I am only
attracted to men with faces," he told me. Their messages cut directly to
sexual preferences and physical assets with no regard for conversational
norms. They simply ignored his nonsexual inquiries.

For him, sexuality is not an illicit transaction. The norms of online interaction strike him as absurd and often alienating. He wants his memories of
sexual encounters to be about pleasure, joy, connection, and fraternity, not
regrets.

He is comfortable with his sexuality and lives in a city where he will run
into the people he meets online or at least occupy the same spaces, such as
a gym, café, or club. He doesn't forget "shitty" online behavior.

Due to his griping, his friends started to refer to him as Miss Manners.
Running with that theme, he started posting many of his exchanges from
Grindr and other sexual media to his Tumblr blog, *Tart Response*. The content resonated with his friends and their friends, so he now curates their
screenshots along with his own.

Take the exchange, posted to *Tart Response*, that ensued after Curtis
wrote to a guy based on his profile picture:

Guy: "I'm only interested in a fuck dude."

Guy: "Sorry just being honest."

Curtis: "You mean no wedding or poodles?"

Guy's response immediately slammed shut the door to any kind of on-going connection. Curtis responded with a joke, calling out the assumption the guy had made about the level of intimacy and commitment he was seeking. Curtis explained to me that he understands that a "fuck can just be a fuck," but he wants even the briefest sexual encounters to be with people he likes.

Curtis may not have been after a wedding or poodles, but he did want some kind of human connection. He explores digital intimacy through joking responses in his online exchanges and by sharing them on *Tart Response*.[8] He receives an implicit kind of support from his Tumblr community. In response to this particular post, his readers didn't commiserate with statements like "I'm sorry you had to go through that" or "You must feel awful." Instead, their empathy took the form of head nodding and humor; his friends shared the joke and laughed with him about it. The insults on Grindr occur in one-on-one, presumably private interactions. By posting them, Curtis broadens the audience. He doesn't suffer alone.

He likens sharing these kinds of exchanges on his blog to calling some-one out for jumping a line. Others in the line support the whistle-blower, even if it is just in their facial expression and posture; there is bonding through shared disapproval. Similar bonding occurred when he worked in customer service: he and other agents would compare notes of outrageous customer complaints and unreasonable demands. As with his blog, satirizing the rudeness that played out in these one-on-one interactions permitted a connection with a broader community.

Tart Response works in part because it is funny. It points to absurdities and turns the tables on unacceptable communication. Like Bye Felipe, it uses humor to foster collectivity among those in a similar predicament.

How We Click

Online dating has become commonplace. According to a 2016 Pew Research Center report, at least 15 percent of US adults have used online dating or dating apps, and 41 percent know someone who does.[9] Most of those surveyed agreed with the idea that online dating is a good way to meet people. This

report notes a threefold rise in online dating among those who are eighteen to twenty-four years old (attributed to the availability of mobile dating
apps), and a twofold increase in use among those fifty-five to sixty-five,
who tend to sign on through websites. To many readers, these numbers will
sound low.

Online dating is popular for good reasons. As previous research has
shown, it does in fact enable people to meet. Those meetings include
hookups, of course, but also extend beyond. More than one-third of marriages in the United States now begin online.[10] Among heterosexual couples,
relationships that start online proceed to marriage more quickly than those
started off-line.[11] These relationships are a bit more enduring than those
initiated off-line, and among those who remain married, those who met
online reported slightly higher marital satisfaction.[12]

The rise in online dating coincides with an increase in interracial marriages, and some suspect a causal link. Cornell researchers Josue Ortega and
Philipp Hergovich traced surges in interracial marriage to the arrival of Match
.com in 1995, and other nexus points, such as the 2012 launch of Tinder,
which has approximately fifty million users and generates twelve million
matches a day.[13] At odds with this view are the blatant racial biases that
persist on dating sites, particularly against Asian men and black women.[14]
Navigating this racism requires resilience, as explored in the blog *Least Desirable*.[15] Despite their problems, these apps are becoming a default way to meet
romantic partners.

The examples in this chapter hint at another reason for using online
dating and hookup apps: these individuals were trying to get more perspective on themselves. Their motivations were largely related to identity, even
in cases when the apps also helped them meet people. Some found support
and personal validation by combining screenshots from these apps with
social media. In other examples, individuals used hookup apps as a form
of medicine, dosing themselves to navigate challenging transitions or cope
with breakups. They weren't escaping these life challenges but repurposing
hookup apps to navigate them. Their strategies worked to varying degrees.
The emotional lift of new matches was typically brief, but in some cases it
prompted other actions and conversations that had more persistent positive effects. Integrating dating apps with larger forums (specifically, sharing
screenshots from Tinder and Grindr on Instagram and Tumblr) effectively

elicited support from the individual's community. Those who focused on accruing as many matches as possible felt fleeting victory, even though this approach was unlikely to yield enduring validation. Others, like the man who explored his sexual orientation and social life around the world through Grindr, used these apps in ways that were profoundly meaningful.

Moving beyond the hookup, dating apps are being adapted for emotional first-aid, social support, and self-exploration.

We continually rework the way we present ourselves as we grapple with new roles and ideals. Choices about which images and words we use to put ourselves forward are part of how we explore these role shifts. And it is through these expanded roles and perspectives that growth can occur, throughout life. As Kenneth Gergen argues in *The Saturated Self*, the pursuit of a "true" or "core" self is misguided.[1] We have not one self but many. In *Life on the Screen*, Sherry Turkle advanced this idea, defining the self as a distributed system, simultaneously playing many roles.[2]

We expand our life stories, or identity narratives, to accommodate these multiple selves. In a form of self-reflection, we observe the conflicts between these selves over time. The life story isn't a passive recap of what has happened to us but rather an active process of integration and analysis that shapes future experience. Psychologist Dan McAdams, who has conducted extensive research on identity narratives, explains that to be helpful, our narratives must have complexity. They need to accommodate differences in the way that we represent ourselves, either as a function of context (e.g., that one is surly with family but cheerful with friends) or time (e.g., that one was raised Catholic but later identified as a Buddhist).[3] Examining our life stories can help us break out of restricting themes, opening up new ways of thinking about ourselves and the future.[4]

Social media and other technologies have become integrated in our identity narratives. We present some aspects of ourselves unintentionally; for example, cues to personality appear in "behavioral residue"—the traces we leave in our physical and online environments, such as an unmade bed or terse responses to comments about a photo. We also intentionally convey how we want to be seen through "identity claims," such as a prominently displayed book, or affiliation with contacts online.[5] During an identity transition, such as the gender changes underway for some of the individuals below, these acts of self-presentation online become more deliberate and charged.

In this context, technology can be thought of as a transitional object: something we attach to as we take risks. In childhood, the transitional object speaks to both connection and a growing awareness of one's separateness.[6] We seek transitional objects throughout life. Just as the toddler clutches a blanket as she risks steps apart from the parent, we clutch our mobile devices as we venture out alone, perhaps seeking the security of our social networks. Instead of recoiling in fear, we dare to look outward and plot a next step. These leaps prompt shifts in our life stories; they allow us to expand ourselves.

This close attachment to technology, like other close relationships, eventually gets fractured. Our feelings about a service, device, or data can all change over time. Just as a child negotiates a parent's boundaries and limitations, or as one spots imperfections in a formerly idealized romantic partner, we confront the limitations of our devices, services, and apps. We realize they are not perfect; they are, in object relations parlance, "good enough." Fortunately, disillusionment is what spurs growth. This disappointment with technology invites us to bend it toward our own ends. Our focus shifts from the technology itself to larger goals and our relationships with others.

There may be benefits to reflecting on our attachments to technologies. Thinking about what we are seeking from them may shed light on what we are striving for in our human relationships. Then, we can examine whether we are using technology in ways that support our interpersonal objectives. This line of thinking is an extension of attachment theory and related therapy models that examine relationship styles, particularly how early relationships with parental figures shape our relationships as adults.[7] In the same way, it makes sense to examine our relations with the devices and data that become so intertwined with our identities and interactions.

The way that we attach to technology is shifting, though. In 2005, Turkle captured the high drama experienced by a woman whose Palm Pilot, a device that only some readers will remember, lost its charge. "When my Palm crashed, it was like a death. ... I thought I had lost my mind."[8] The smartphone quickly subsumed the Palm Pilot and other personal digital assistants and obviously became a close object of attachment. For a time, the phone itself felt critical for our social and professional viability. It was as if our friends and colleagues lived within the physical device, and the phone reassured us that we were not alone. But this attachment is in flux. The current sentiment may be reflected in a scene from the TV series *The Mindy Project* (the "Hot Mess Time Machine" episode, aired in 2017), where Mindy drops her phone into a cup of soda and shrugs. It's just a phone. Similarly, after my colleague dropped her phone down an elevator shaft, she described "mild inconvenience" and a short, welcome break from news and social media. She was not broken over her shattered phone. As more of our data moves into the cloud, our attachment to devices is likely to become even less intense. There is typically some financial pain to replacing these devices, but they are becoming easily replaceable commodities. They are weaker signifiers of our social selves than they once were.

Our identities are likewise malleable and resilient. Using technology creatively, we can change the stories we tell about ourselves.

Masking Emotional Effort

In an earlier chapter we met Paula, the woman who was using Tinder to fine-tune her criteria for finding a culturally compatible match. Her parents had broken with the Indian tradition of arranged marriages and instead married for love. But their relationship was a rocky one. Some of Paula's earliest memories are of her parents fighting. Harsh accusations and screamed regrets about ever having started a family permeated the walls of their home. Paula could not escape these situations as a child growing up in Bahrain, and she has never spoken about them with her siblings or parents. These arguments were the only real emotional expression in her home. It was only when watching white families on TV that she saw emotional vulnerability play out in a positive way—fictions that were worlds apart from her own experiences.

As a teen and now as an adult, Paula tries to avoid conflict or any interaction where she might come across as vulnerable. She works at presenting a consistently cool exterior, even when she is angry or sad. She plans out her interactions and works equally hard at effectively concealing this effort.

One way Paula has concealed her effort is by composing text message replies in the iPhone Notes app rather than using the iMessage interface. Once she's crafted the perfect nonchalant reply, she pastes it from the Notes app into the messaging window. The recipient never sees the bouncing dots that iMessage displays when someone in a chat is typing a message. That signal would betray how much time and effort Paula was putting into her simple text response.

Paula told me about one time that she needed to hide these bouncing dots. She had brought a friend to a party. On the way, Paula mentioned that she and the host had previously dated. The next morning, her friend texted somewhat apologetically to inform Paula that she had hooked up with her ex. Paula was upset by what felt like a violation of an implicit code, but she wasn't sure that her reaction was justified. She struggled with her response to the text. Her first take, crafted in the Notes app, was something like "I'm a little uncomfortable with this, but OK." After reflecting on it and consulting a friend, she reworked the message to something far more laissez-faire: "Yeah, I figured haha. No worries!"

Although hours of thinking went into this response, her friend saw only the final, seemingly easygoing result. Paula would have been embarrassed if her friend had seen the bouncing dots, a sign that there were in fact worries and that Paula's seeming indifference was actually the result of considerable thought and effort.

This minor subterfuge, a suppression of an unwanted signal, works. But it is not without a cost. In this case, Paula didn't think the friendship was strong enough to merit emotional openness and honest confrontation. It is possible, though, that the technique makes it too easy for her to put relationships into that category. The nonchalant self that she is constructing may miss opportunities for taking the emotional risks that strengthen trust.

An Assertive Self

Alyssa is a highly energetic college student studying physics. She has also volunteered for leadership roles to coordinate events on campus related to specific social causes. Despite her desire to lead, she does not feel fully confident directing others.

Shortly before we talked, she was pulling together an event that required volunteers and space from organizations on campus. This meant recruiting and overseeing students with whom she did not already have a relationship.

One of those new volunteers, Ed, wrote an email demanding more recognition for his work as well as special privileges beyond those granted to the rest of the team. His note to her was one long angry rant, with no paragraph breaks. She politely responded, over several short paragraphs, that she could not meet his requests; it would be unfair to the other volunteers. They went back and forth a couple of times. She could see that as their conversation progressed, he started mirroring her writing style: he started using paragraph breaks, and his tone became less aggressive.

Alyssa observed this pattern through a feedback tool that she agreed to use as part of our study: a Gmail extension intended to help users mirror other people.[9] Rather than monitoring her own communication with this feedback, though, Alyssa used it to scrutinize Ed's language and his mirroring of her. She could see that his counts of positive words and other metrics were getting more similar to hers over the course of their exchanges.

Alyssa took Ed's mirroring as an indication that he was trying to work with her. This helped soften her opinion of Ed's character, and it also affected her view of herself. Noticing his mirroring made her feel respected and confident. She could see herself, through his eyes, as a leader. She then began to fulfill this image of herself by speaking and acting more assertively.

The feedback from her device helped her more fully inhabit a self that she had tried on with some trepidation.

Appearing Interested

Kara is anxious about time. She traced her concerns about efficiency to the strict boarding schools that she attended in Beijing, where she had to account for every moment. When we spoke, she had a heavy course load and was trying to shift to a competitive major. She had wanted to be in a relationship for a while and suspected that her attempts had failed because she was unwilling to spend enough time with the other person. It occurred to her that maybe she just needed to get smarter about how she used her time and more communicative about her availability. But when she and a boyfriend tried a shared Google calendar, she felt restricted from dynamically adjusting her schedule as the day progressed. She felt that she had to stick with what was on the calendar, or waste time changing the calendar and explaining why she did so. She felt trapped.

It was worse with acquaintances with whom she was pretty sure would not become close friends. Kara didn't want to waste time in these interactions, but she also worried about hurting their feelings. She wanted to give the impression that she was excited to see them and enthralled in whatever conversation ensued. So she developed an escape strategy: when she ran into someone she felt obliged to speak with, she set a timer on her phone for ninety seconds or a bit longer. She would start talking and then feign disappointment that she had an obligation and had to dash.

The system worked but wasn't perfect. If she was seen setting the timer, appearances would unravel. Her lack of interest in the conversation and lack of investment in the relationship would be revealed, and she might be seen as untrustworthy.

That's not how Kara sees herself in these situations. Kara wants to invest in relationships, but not any old conversation. She sets a high bar. With the timer, she sought to protect the feelings of people who would not meet her standards.

Kara has been using WeChat, a popular instant messaging tool, in an effort to reestablish meaningful friendships. She doesn't use Facebook and is put off by the similar Moments feature in WeChat. She cherishes WeChat's message history, though, seeing it as a chronicle of her friendships. Starting at the bottom of her message history, she recently picked up conversations with old friends, some of whom she went to high school with in Beijing. In these online exchanges, she signals that she values the friendships while opening up the communication on a channel that makes it far easier for her to engage at her own pace. Freed of the subterfuge required by face-to-face encounters, she can try out a different social self. Online, she slows down and can express her appreciation of the relationship.

Our culture puts great emphasis on face-to-face conversations. Those are the real ones, we generally believe. It is in face-to-face contact that we think we see the real person. But there isn't just a real self and a facade; we are continually developing ourselves by trying out different roles and different modes of relating. For Kara, who is bedeviled by a sense that time is slipping by irretrievably, it may only be in situations where she can control time that she is able to reveal and develop her more sociable self.

Conveying Loss and Hope in Metaphor

Marlena wandered alone through an evening arts event. In one installation, an array of Instagram photos, those of other attendees, were projected on the wall in front of her. She decided to share one of her own. She selected a photo from her phone gallery, and, when uploading it to Instagram, captioned it with the words "death" and "miserable," along with the event name. A moment later her photo appeared with the others on the large display wall. It was a hauntingly beautiful image of a skeleton. The image stood out from the other celebratory photographs on the wall. Marlena told me that she had previously taken the photo at a museum, but she shared it on that day to convey her recent breakup.

This installation, one that my colleagues and I developed, associated emotions with images shared on social media.[10] Each image was framed by a color based on the sentiment in the image caption. For example, an image with the caption "fantastic" would be tagged pink, whereas one captioned "bored" might be gray. The system also asked people to describe how they felt when seeing images. Using a mood map that appeared next to each image, viewers could adjust the classification of their own photo or express how they felt about someone else's photo.

Marlena's skeleton image was framed with a sad gray because the words in her caption—"death" and "miserable"—were categorized as negative and low energy.

After looking at the image and talking about it, she decided to emotionally reframe it. She explained to me that it was a painful loss, but that she was optimistic, not sad. She walked up to the wall, and through a gestural interface, dialed the frame from gray to pink.

Marlena corrected the software program's interpretation of her words. The classification of her image and caption as negative was logical, but it didn't capture her full emotional experience. Many images of loss, or even despair, are beautiful and uplifting. Our emotional responses to art and events in our own lives are multifaceted. It is possible to be optimistic, as Marlene was, while acknowledging a loss. While sentiment analysis can classify the emotional tone of some of the words that we use, it is not always able to understand mixed feelings or how feelings change over time. It is up to us, as Marlena shows, to use our own devices—in this case, permutations of language, images, and color—to convey the textures of what we've experienced and where we are headed.

Another woman at this same event posted a glamorous picture of herself and tagged it with the word "miserable." As she laughed with her friends, it was clear that she was training them and her Instagram followers online to appreciate her emotional complexity. Like Marlena, this woman was expressing a more complex relationship between her image, feelings, and social self than typically comes across on social media. Her laughter ran counter to the sultry photo and dark caption, creating tension and a warning that she might not be as cheerful as her followers assume. This could all be interpreted as a mask: laughter and glamour covering her real misery underneath. Instead, I see it as a textured performance of a social self.

Such multilayered performances of self aren't limited to these two individuals of course. It is worth challenging ourselves to express and grapple with complexity. And it might get easier to do so. Sensors and algorithms will become more adept at handling mixed feelings (e.g., miserably happy) and the play of multiple meanings—such as a skeleton labeled "death" and "miserable" that is meant as a symbol of hope. Others won't always understand complex expression, though. Take the funeral selfie, which draws harsh critique, even though it is a way of connecting during a significant loss.[11] The more nuanced expression in selfies and previous forms of self-portraiture, particularly those by women, have frequently been overlooked and dismissed as forms of vanity.[12] Whether or not emotional complexities immediately register with all of one's followers, there is still value in trying to express them. Articulating nuances associated with poignancy, bittersweet wins, or the anxiety that accompanies an opportunity might draw in others who have had similar experiences. And as discussed above, articulating emotions with specificity seems to help us respond to stress in thoughtful, constructive ways rather than with knee-jerk reactions.[13] In addition to these benefits, textured emotional expression allows us to have more interesting conversations with others and ourselves.

Selfie Defense

Changes in self-representation on social media take on a heightened urgency for people who have transitioned or are gender fluid. In the next several examples, we meet individuals who have made profound changes to the way they see themselves and present themselves to the world—changes that go far beyond the selection of a checkbox for preferred gender. They grapple with reshaping their identities and portraying that evolution online.

Sophie has had a long and challenging transition from male to female. Early on in this process, she spent a week visiting a city known for its openness to the LGBTQ community. But even there, she immediately confronted harassment. As she walked down a busy street, a man shouted an insulting sexual comment to her. She felt fetishized for being trans. She wanted to hide under the covers. Instead, she "stared down the shame," posting a selfie in which she wore bright lipstick, a polka dot blouse, and flared eyeliner. She posted this to her public Facebook page with the caption "femme resistance."

It's easy to lump together all selfies as a narcissistic quest for approval. Far from that, Sophie's selfie challenges transphobia and builds community. She celebrates the very aspects of her appearance that were targeted by the harasser. The image and caption elicited support from others in her queer and trans communities. "I've been there," one friend commented. Sophie is tolerant of selfies that are intended to elicit "you look great!" comments, but that's not what hers was about.

Sophie has used the internet for creative expression for almost as long as she can remember. As a child, she was the first to understand and fix the family computer. In her teens, she found freedom in blogging on LiveJournal, posting mash-up videos to YouTube, and reading in the local zine library about others who were exploring gender identity. Participation on Ask

MetaFilter and Reddit enabled conversations and connections that were not possible off-line, whether on the basketball court or in her social life as a teenage boy.

Through this online exploration along with a more painful and circuitous one off-line, Sophie delved into her sexual and gender identity. Sophie came out as bisexual at age fifteen and then as queer in sexual orientation at age twenty. In her twenties, she openly examined her gender, identifying briefly and mostly online as genderqueer, and then openly coming out as a trans woman in 2014, at the age of twenty-five.

Like others in the trans community, Sophie has actively managed her Facebook profile to reduce the chances of being outed. Even though she is openly trans, she wants to be in control of her own exposure. In contrast to broad sweeping approaches that are commonly used, such as deleting one's entire Facebook account to start anew, Sophie edits her account in a more fine-grained way.[14] She continually curates, partitioning content that is only appropriate for her private support groups on Facebook (such as insurance and legal concerns) from content that she wants to share publicly, and thinks carefully before sharing or removing content from her profile. She deleted her previous profile pictures but retained the pictures that she had been tagged in. She didn't want to delete these traces of her warmth and joy.

Sophie recently posted a double selfie: one recent image of herself smiling with friends, and one from years ago, in which she was driving and laughing with her girlfriend at the time. Her caption included the words "more continuity than discontinuity." While some of her friends completely reject their "boy-mode" pasts and feel that life began after their transitions, Sophie continually negotiates her history. She retains just this one photo of herself as a boy on Facebook. Taken during a trip that solidified her desire to transition, the photo memorializes both that awakening and a since-deceased friend who supportively commented on the photo. Sophie doesn't share her previous "dead name" with others but deliberately exposes the parts of her past that she wants to carry forward. Her curation invites us to imagine something like a Photoshop for Facebook profiles: some nuanced way of editing our online self-presentation so that what we show the world is most aligned with what we value and how we see ourselves.

Sophie adapts social media to convey her evolving identity. Facebook requests a singular "authentic" self. It wants users to portray who they "really are." But identities change and multiply, sometimes dramatically. Even when identity evolution is less dramatic than a gender transition, some level of creativity and care is needed to retain continuity while allowing for change.

Shedding Syllables

Like Sophie, Alex signaled a change in gender identity. Her transition was the rejection of a fixed gender. By announcing on Facebook that she was changing her name from Alexandra to Alex, she opened up desired fluidity in how she would present herself and be seen by others.

The extra syllables in Alexandra had come to feel dissonant to her. She associated them with ideas about gender and heterosexual norms that she no longer accepted, and no longer needed to feel accepted herself. Alex now goes by "she" in some situations and "they" in others.

Her new name is in fact what she had been called as a child. Those resonances, however, are mixed. Growing up, there wasn't a lot of encouragement for expressing feelings or vulnerability. Alex's Korean parents valued toughness, pursuit of success, and self-sufficiency. In this way, one might say that they raised Alex as a boy. But small-town conservatism was also in effect: Alex was not listened to as a son might have been. She received little nurturing or acknowledgment but did have a fierce role model in her father: though they have little else in common, she shares his ambition.

Alex's parents arrived in the United States traumatized by war. Her father is in business for himself. Her mother worked out her anguish on her body, following a grueling athletic routine throughout her life. Alex noted "a thin line between self-care and self-abuse" among all the women in her family. Seeking security and healing, Alex married shortly after college. Getting married brought validation: her parents saw her as doing well. Still, when she visited, they talked to her husband rather than to her and constantly asked when she would have a child.

Realizing that she was headed down a path toward motherhood that she had never wanted, Alex left her marriage after a couple of years. She endured a long silent treatment from her father as a result, but for Alex this was an exciting time to assert her independence and make her own choices. She began experimenting both artistically and socially, dating men, women, and individuals who were gender nonconforming.

Alex also began experimenting with how she dressed. She has long viewed her clothing as a costume and is amused by the assumptions that people make based on something as simple a shirt. "If I wear prints," she laughed, "people think I am Japanese." She used clothing to play with her identity in multiple dimensions, dressing as a Chinese woman from the 1950s on one day and on the next day, a Japanese man from the sixties.

Alex enjoys the fluidity that this gender play gives her socially and in her work. Her gender presentation at any given moment depends on social relationships and context. Within lesbian and queer social life, Alex feels as alienated by the exaggerated roles of butch and femme as she does by the traditional roles tied to one's biological sex. She doesn't want to be categorized. In her current relationship, she doesn't want to be locked in as her girlfriend's girlfriend.

And even though she appreciates the third-person plural and uses it for others, she doesn't want to be stuck there either. In part, she doesn't want people to get too nervous about what to call her. What she wants is respect for her choices. "You can get the pronoun right and still treat me like shit," she laughs. She traces this desire for fluidity to Buddhism and a transcendence of gender binaries.

Since leaving her marriage and her identification solely as a woman, she has become much more self-defining in her work, completing major projects independently and receiving several prestigious grants. The most important shift in her life, as she looks back, has less to do with gender than it does with agency. She is no longer in a supporting role.

To separate herself from a fixed gender, Alex experiments with the images as well as the names associated with her social media profiles. She recently changed her profile picture from a full-body image to one of the top of her head. The artfully framed image of shiny dark hair jumping playfully upward against a white backdrop signals that Alex would like to be seen, but not immediately as either a man or woman.

Alex said that were she to ever transition, she would probably wait until her father died. She doesn't want to land a gender bomb at the end of his life, when he is coping with isolation and despair. Her mother is quite ill and can no longer speak; there will be no disclosure to her. Every time that Alex sees her father, he asks when she will marry the physician son of his friend and when she will have a child. She laughs and tries in little ways to break through the barrier. "Dad, can you really imagine that?" she challenges him. "Don't you see that you raised me to be more like you?"

With social media, Alex takes explicit control over the most obvious gender signals. By playing with her name and photo online, she lets people know where she is at the moment and also that wherever she is now may not be where she'll be later. She exercises similar control offline by dressing in costume; her play with clothing signals that she is playing with how she wants to be seen. But she can't change her real-world persona as quickly as she can change her virtual one.

Online and off, her appropriation of social media and clothing resonates with the audiences she is trying to reach. Her friends, many of whom are gender queer, can dial into these cues and support her in her fluid play. Her parents can't and don't.

Signing Off

When Pat sends an email, the automatic signature doesn't offer a professional title or an inspirational quote. Instead it's their name, contact information, and on a separate line, "they/them/theirs." In this way, the email conversation partners are informed that Pat is genderqueer.

A standard email signature conveys contact information and sometimes instructions for addressing the other person: "MD" invites us to address the person as "Dr.," and a long professional title often signals a desire for status recognition. Pat's sign-off gives analogous guidance. Unlike the typical signature, though, it asks the reader to pause for a moment to reconsider gender.

Pat never felt like a girl or boy growing up, and has never had a desire to be a man or woman. Pat lives between these poles, feeling more like neither than both. As a kid, Pat rejected being referred to as a boy or girl and settled unsatisfactorily with being called a tomboy. In college, they found others who felt the same and became interested in gender fluidity and queerness. Pronouns were an important means of exploring and communicating gender fluidity within this community. To ease into presenting as queer, Pat asked several close friends to use "they" and "them," or "ze" and "hir," when referring to them. Over time, Pat has become more comfortable asserting gender fluidity by requesting this plural pronoun use of everyone, not just close friends.

Not wanting to make assumptions about others, Pat often asks, "What are your pronouns?" early on in conversation. There are trials every day, but

Pat tries to view these as teaching moments, where patient explanations might ripple beyond the immediate interaction. To one friend, it helped to explain the grammar by replying, "Imagine that there are two of me." The idea of plural selves resonated and helped this friend appreciate the shift to third-person plural pronouns. A group of Pat's close nonbinary friends, all people of color, huddle in Snapchat in a group that they call "the angsty boys." They share jokes, updates, and practical support.

Pat's expression of their preferred pronouns in the usually staid email signature block subverts expectations, creating its own learning moment. It does so at a safe enough distance that readers can reflect on the question without feeling that they've been scolded. Colleagues and friends see this signature frequently—repeated instruction that may help chip away at ingrained pronoun use. Including preferred pronouns in an email signature has become a practice among some work groups, particularly in universities. Pat includes the "they, them, theirs" sign-off in their personal email as well, spreading the influence to people who may be less familiar with these practices and preferred pronoun trends.

Writing is often considered to be more remote and less personal than face-to-face communications. Pat leverages email precisely for its distance. They can educate others without reprimanding or creating a lot of anxiety. And the sharing of personal pronouns, an important issue for Pat, isn't postponed for a later face-to-face conversation. Embedding pronoun preferences in email helps Pat and their correspondents address each other in ways that align with how they want to be seen.

Our Devices, Ourselves

In the examples above, individuals use their devices and social media to inhabit different selves. Curating social media profiles can help us explore our varied identities with some degree of control. By posting pictures of ourselves in different roles, we engage with those varied identities, sometimes cultivating new parts of ourselves.

This was the case with Kathleen, a college student who spent most of her first year in her dorm room, procrastinating and worrying. It's painful for her to look at pictures of herself from that time. They remind her of a self she wanted to leave behind. A possible self, an inspiring concept of who she could become, emerged over that summer.[15] Volunteering at a national park, she met some avid climbers and started joining them on expeditions. Kathleen admired them, in person and online. She confronted the difference between her life up until that point and the adventure depicted in their photos. She sought to become more like them and used Facebook images as a measure of her progress, continually comparing hers to theirs and trying to lower the contrast. When the climbers posted pictures that included her, Kathleen started to see herself in a new way: she too was an adventurer. She started posting her own climbing pictures and sought to post ones that were as compelling as theirs. This required her to go to the high lookout points—a challenge that called for not just bursts of courage and energy but also the discipline to learn how to climb safely.

While she does occasionally post selfies, she far prefers curating images that others tag her in. These vouch for her. They offer social proof, not just an identity claim. These images motivate her to keep adventuring, to keep moving toward her ideal self. She deliberately pursues experiences that will generate these kinds of photos, to reinforce and motivate an openness, physical confidence, and what she calls "stoke."

But pictures aren't the only way to amble between our different selves. Sebastian, a lover of computers, installed Windows software on his preferred computer, which is a Mac, using virtualization software so that he can toggle between his work self and his home self with ease. He uses the Mac to edit videos with his nephews that he wouldn't want to be visible to colleagues. At work, he is a Windows guy, not a high-maintenance Mac user, but when huddled over the laptop in Mac mode with his nephew, they both feel like movie directors.

Sebastian is not alone in wanting to keep his professional, social, and family identities separate. Microsoft researcher danah boyd has explored the downsides of worlds colliding, the phenomenon she calls "context collapse."[16] In *It's Complicated*, boyd shows how the insistence on a single identity in tools like Facebook can harm those who need or prefer to keep their social subgroups separate from one another.[17] In one instance, boyd describes being consulted by an Ivy League college admissions officer who was concerned about gang-affiliated communication on a candidate's social media presence (particularly since this candidate had written about wanting to escape gang life in his application essay). This gang signaling, boyd pointed out, may have been a strategy for ensuring survival where he lived in South Central Los Angeles rather than a representation of the student's "true self." To partition audiences, many teens use multiple accounts on Instagram and Twitter or write updates in a coded style that only their closest friends will understand, a strategy Alice Marwick and boyd call "social steganography."[18]

You may have your own systems for preventing your worlds from colliding. It may be coded communication that only your intended audience understands, the use of different apps or multiple accounts for different relationships, or something like Sebastian's system of running Mac and Windows operating systems on a single machine, which enforces a separation of his family and professional identity.

These efforts are warranted. We all have multiple roles involving different relationships and codes of conduct. We may experience ourselves fundamentally differently across these roles. The style that works in one role may violate the expectations associated with another; it is easy to imagine aggressive negotiation tactics going over better in some work environments than in intimate relationships, for example. These selves change and multiply across our lives, sometimes quickly. And as our selves evolve, so do the possible conflicts among them. The line between personal and professional identity, or between family and friends, may blur over time. So like Sebastian's software, which requires updates, our own partitioning strategies may need frequent tuning.

A Self in Conversation

A man in his forties, apparently homeless, sat on the street corner. He held up a twelve-inch cracked mirror to his face as he spoke back to a boom box blasting talk radio.

A couple walked by. "Clearly off his meds," she said. "Off his rocker," the man replied.

The man's dialogue with the splintered image of himself and the radio could be an image in a TED talk portrayal of the new universe of selfie sticks and tweets that no one reads—a universe in which we think we are being social when we are in fact more adrift and alone than ever. With this selfie machine, assembled from discarded objects, he dramatically performs the perils of social media participation. He is speaking, but no one is listening.

We could, though, take it differently. We so want to be in conversation that we will do what we can to piece together the tools that let us project ourselves outward. It is a basic human need. Even those lucky enough to spend their days in face-to-face conversation use strategies, like the ones described in this book, to expand their dialogue with themselves and others. We share with this man the deep desire to listen and be heard.

As the "sharing economy" challenges traditional businesses, the dark side of these businesses has justifiably dominated the news. Daily headlines report on how companies exploit their workers because they are not exactly employees, manipulating them into working longer stretches of time than they intended.[1] We have also read news of dangerous encounters with unlicensed, barely vetted, strangers.

If, however, you were to listen to Brian Chesky, the founder of Airbnb, you would hear an idealistic vision in which citizens open their doors to visitors from around the world, inviting them into their homes, neighborhoods, and social lives. The Airbnb corporate mission statement, "Belong Anywhere," reflects the company's broad sense of its role.[2] Far from just offering lodging, the company apparently strives to generate acceptance, trust, and global connectedness.

The hosts and users of services such as Airbnb clearly get something out of them, often beyond a financial benefit. Whether it is a weekend in an apartment or ride from an airport, a place and a time are shared between strangers. Conversations arise more readily than they might otherwise, in part because the customers understand that the service is being provided by someone who is probably not doing it full time and has a life beyond the confines of the interaction. When was the last time you talked for twenty minutes with the hotel manager about her favorite neighborhood cafés or with a cab driver about the life he left behind in the country of his birth?

The oft-maligned sharing economy not only occasions conversations but invites them. There's an empathy that arises from the feeling that roles could be reversed—the host might one day be the guest, or the driver might one day be the passenger. This, along with the constraints of the interactions— visits and rides are finite—may engender a particular kind of openness. The conversations provide opportunities for listening and reworking one's life story. The connections also play out in a less explicit way: we absorb the other person as we get in their car, drink from their teacups, or take their preferred path through town. Their routines can get under our skin and become part of who we feel we are.

Some of these stories, about brief connections, were derived from my own brief conversations as a passenger. Others are from people I've interviewed. These are not the headline accounts of a marriage between a host and guest, or a driver's mad dash to a hospital for an expectant mother. These stories are about the seemingly banal but psychologically meaningful connections that we have with people we will likely never see again. It is also about how the accumulation of those connections influences our identity.

The Long Ride Home

It's hard to become a taxi driver. You have to know the city like the back of your hand, pass a rigorous test, and invest in a license and medallion. It's a profession that inevitably becomes part of the taxi driver's identity.

But when you're in an Uber or Lyft, you probably guess that the driver sees himself or herself as more than a driver. To probe on this, passengers often ask, "Do you do this full time? Do you have another job?" The conversations and interactions are frequently different than they are in taxis, where roles are more fixed. It may be easier to relate to the other person and even imagine being in their role. Whether you are the driver or passenger, there is an opportunity to engage with attentiveness and empathy. There is no threat of an ongoing burden: the relationship ends at the destination.

Sometimes, though, Uber's technology, which reflects the company's policies, creates rules that individuals prefer to break for the sake of human connection. Take the story I heard from a driver who picked me up at JFK.

On the day of the New York City Marathon, an Uber driver picked up two women who had been drinking heavily. One vomited in his car shortly after she got in—not a great start. The woman who hailed the ride wanted to take her drunken friend to her home in Connecticut and then return to Manhattan. Due to surge pricing on this busy holiday, the round-trip fare would be over $1,000. After incurring the $540 fare to Connecticut, the remaining passenger announced that she could not afford to return and would instead stay with her friend.

The driver would have lowered the fare to Connecticut if he could have because the surge pricing seemed unfair: it applied to where the ride started and probably passed within half an hour. But the Uber app that drivers use doesn't allow this sort of discounting. And since he was going back to New York City anyway, charging another $500 seemed excessive. So he navigated around the limitations of the Uber app simply by not entering the return trip as a paid trip and drove the woman home for free.

During the long drive to Connecticut, he and the woman had talked. He liked her. His concern transferred from the fare to the welfare of his passengers. He had long stopped caring about the vomit in his car.

With the Uber app, users don't choose their drivers or passengers by reading through a profile or scanning through pictures. The app masks the identity of those asking for rides as well as their destinations. It likewise masks the Uber driver's name and photo from the fare requesting a ride. Although drivers can, with a bit of work, obtain this information and some "elite" passengers may have more visibility into their drivers, most know little about the person with whom they will be riding. This technologically encoded decision throws together people who may have little in common, as was the case here. The driver was struggling to take care of his family; the passenger was single and independent. He was working on the Fourth of July; she was getting drunk. She was bummed out about the $1,000 fare, but this would not threaten next week's groceries.

A relationship is facilitated by the constraints of the ride; it is a bounded experience, with a beginning and known end. There are ready-made things to talk about, such as traffic and areas of town. The driver, too, often becomes the subject of the conversation. Strong connections can form among people who otherwise would probably not have met, and, if they had met, probably would not have talked.

In this case, a trip that began with a blind pairing of driver and passenger ended with the driver blinding the technology so that he could do the right thing for the person he was getting to know.

Uber Elevator Pitch

Our life story is what we tell others and ourselves about our past, present, and future.[3] There is value in examining it, finding ruts, and thinking about how we want it to evolve. It can be helpful to explore our life stories with strangers as well as with those we hold close. And for that, the sharing economy holds promise.

On one of my rides home from the Seattle airport, my Uber driver, Araya, summed up his life: "You can do anything if you are a survivor."

This wasn't exactly what I expected in response to my question of how long he'd been driving, but it was clearly the story he wanted to tell and the self he wanted to present. Consistent with this presentation was his clean, well-maintained, aging car and his slightly formal, dignified manner. I was curious about his story, and he spent the rest of the ride telling it. After deportation from Ethiopia to Eritrea in the early 1990s, Araya came to the United States to seek asylum and visit his nephew. He stayed, successfully obtaining asylum for himself and his family members. Since then, he has worked steadily to reestablish himself.

Araya was trained as an accountant in Eritrea, but he had to take other jobs in the United States because of certification requirements. First, he worked as a gas attendant on the night shift, and then he started driving for a private car service. The pay was better, but it put his time at the disposal of the business. The car service's central dispatcher notified him whenever there was a passenger to be picked up, and he was expected to respond immediately. It was a demand-driven, unpredictable schedule.

He was definitely surviving and was grateful for it. But it was only when Uber partnered with the car service company that he could finally take a step back and think about his goals. At that point, Araya set his mind to resuming his former career as an accountant. He realized that he would need to devote a great deal of time to prepare for his US certified public accountant exam. Taking advantage of Uber's decentralized system, he could set his own hours for driving, balancing his need to study with his need to earn a living. Through this small technical change introduced to optimize Uber's business—a decentralized app versus a central dispatcher—he was able to resume control over his life.

He reunited his family, is reestablishing his professional life as an accountant, and is relating to passengers with confidence. And his view of himself and his prospects changed. Before network-based technology allowed him to regain control over his schedule, he was definitely a survivor, but at any point his life could have been derailed by the loss of his job. With this technology, he was able to envision his future.

That is in part because Araya now has more control over his story. It is a story of redemption, from hardship to success, achieved by hard work and seizing timely opportunities. His story abounds with what psychologist Dan McAdams calls commitment narratives: he emphasizes how much he values family and meaningful work, and ties those values to the sacrifices that he has made.[4]

The facts of the story come from what he has been through, including the opportunity opened by the sharing economy. But the arc of Araya's story was, I believe, worked out in conversations with hundreds of passengers who, like me, asked him about himself in a ride that was bounded by its here and there. These short rides were the medium for reworking his life story and, in some important way, his future.

The Company of Strangers

Following a midlife divorce, Miranda was bereft emotionally, financially, and socially. After a period of dejection, she considered all the things that she could do to bring in an income: dog walking, tutoring, assisting older adults, and Airbnb hosting. She has been doing all of these since then.

She often hosts several Airbnb guests a night, relishing the conversation and vicariously enjoying their travel. Several of her guests have become close friends. She emphasized that she would never want a roommate, but loves the company of strangers in her home. Miranda is remarkably warm and frequently drives her guests around during their stay. It's clear from the way she described her humble, crowded home that she's making a living from her personal generosity rather than a well-appointed property. She becomes a mother to her guests for the duration of their stay. Her nurturing manner is something entirely different from the polite formality of a Hyatt concierge.

Of course, finances motivate most Airbnb activity. Airbnb rentals are generally less expensive than hotels. And for the host, it is often a source of significant income. Miranda is one of many people who have found financial grounding through hosting. But shelter is not the only or even the primary offering. In many cases, certainly in Miranda's, it's banter, nurturing, friendship, and other human qualities that appear to drive the business.

The nature of the Airbnb transaction may make conversation easier for people who are not as social as Miranda. It can also help shape the conversation, offering what Jyri Engeström calls a "social object": an entity that provokes a conversations, and around which relationships form.[5] If you walk a cute puppy down a street, it becomes the object of attention, affection, and conversation. For a user of Airbnb, the place (both the home and surrounding location) is similarly likely to be a source of bonding. The intimacy of staying in someone else's space may invite leaps over small talk to more sustaining conversation.

As a host, Miranda diversifies her social life and enjoys warm interaction with new people each day without the dread of negotiating her space with the same person over a long period time. Her sense of self broadens as she absorbs the adventures taken by her guests. Although she rarely leaves the Pacific Northwest, she feels like an international traveler.

Like my Uber driver, Miranda used the sharing economy to change the arc of her life story from upheaval to survival.

Fountain of Youth

Joseph's taxi dropped him in the Balat district of Istanbul. The apartment appeared even sketchier from the street than it looked online. He reassured himself that in a couple of days, he'd be shifting to a room he had rented in the fashionable Cihangir district. In any case, he hadn't selected this place thinking that it would be upscale. It was the profile of the host, Danyal, that had drawn him in. Danyal struck him as someone who really wanted to meet people and help them get to know Istanbul. Joseph wanted to be welcomed more than he wanted to be pampered.

From the moment he entered the apartment, he felt comfortable with Danyal. They walked around the neighborhood, shopped, and ended up making dinner together. A day later, Tanzeem, another guest arrived. She was visiting from Paris, also drawn in by Danyal's profile. While Danyal worked, Joseph and Tanzeem followed missions that he had set for them to explore the city. They'd reconvene at night, meeting up at a restaurant or for drinks on the roof. Joseph described their near-instant camaraderie: "We were like the Three Musketeers!"

At the end of the week, Joseph moved to the other location that he had booked in Cihangir, a stylish apartment owned by a gay couple. This should have been perfect, but he missed his friends, and they missed him.

Texts started streaming in: "Lunch?" "Are you having fun?" and ultimately "Come home!" And so he did. Instead of moving on with his plans to visit Epiphysis and Cappadocia, he returned to Danyal's apartment and enjoyed another week with his newfound friends. At some point, Danyal and Tanzeem (both in their early thirties) asked his age. Joseph told them that on this trip, he had turned sixty. In what felt like a celebration of vitality, the next day he and Tanzeem took the ferry to the Princes' Islands and rode bicycles all around, running into an impromptu music jam on their way home.

He had divorced several years earlier, after a relationship of twenty-one years. He and his husband had traveled extensively and had always stayed in nice hotels. Joseph enjoyed this luxury but always felt like a tourist, set apart from the places he visited. He also recalled one solo business trip that required him to circle the globe in ten days. This might have been an opportunity to learn countless local customs and even acquire a few words in a range of languages from friendly new faces. But apart from his work meetings, he spoke to only two people.

For years after his divorce, he did not travel. He had a small freelance business and remained close with his partner's extended family, hosting family dinners and never missing a birthday party, but he felt like he was playing it safe. It was as if he were waiting to be in a couple again before having a life. This was at odds with how he wanted to see himself. Pushing against his comfort zone, he decided to travel again, but this time knew he'd have to do more than just get on a plane. He had to engage with people, and not as a customer. He decided to try Airbnb.

The structure of Airbnb fit with Joseph's social objectives. It brought him into the homes of those who were genuinely excited to meet new people and make them feel at home. It created situations primed for intimate friendships without the normal requirements of mutual friends or shared interests. The warmth that he experienced in this context was nothing like the contrived hospitality in many bed and breakfasts. It was more of a personal connection, facilitated in part by his own openness. As a guest, he tries to avoid building any expectations and frequently reminds himself that he needs to be flexible. The host is allowed to be a control freak, but, as a guest, he knows he is not. Even if a home isn't exactly to his liking, it still feels like a secure home base, a source of comfort after venturing out alone to a different part of the world.

These experiences have sunk under his skin. Some of his friendships with hosts have persisted for years. Danyal recently visited him in the United States, as a friend rather than an Airbnb guest. Signs of other connections live in photographs and artifacts, such as a hydroponic plant that is a perfect resemblance of a host's plant in Puebla, Mexico. The clearest imprints of his Airbnb travel are not the periodic guests or decorative objects of course but his openness, flexibility, and comfort with the unknown.

Some research suggests that people narrow their social interests as they age, but Joseph is moving in the opposite direction, increasingly opening himself up to new experiences and people.[6] That includes opening his own home to Airbnb guests, allowing him to reciprocate the welcome that he has experienced worldwide.

Clean Slates

Belinda, who drives for both Uber and Lyft, prefers Uber not only because it keeps her busier but also because she has found it easier to engage her Uber passengers in conversation. They're "nicer," she said, sharing an opinion that contradicts the common stereotypes of these two services as if it were fact.

That comfort wasn't there for her one night when three adults got into the back seat of her car and began a ménage à trois. Conservative by nature, Belinda was shocked and tried not to look in the rearview mirror.

Several weeks later, two of those same individuals got into her car again. She did not approve of their past behavior, but she smiled and struck up a conversation with them. For Belinda, each ride is a fresh start, a new possibility for finding a friendly connection.

The chances of having the same driver twice are low. But even in the rare case where this happens, Belinda wipes the slate clean and welcomes the repeat passengers as strangers. This opportunity to wipe the slate clean, rarely available to us off-line, is invited by the structure of ride sharing. Drivers accept rides without knowing details about passengers and their destinations, which requires some open-mindedness. And except for passengers granted special status, this blindness is a two-way street. The only way to get a different driver is to cancel the request and begin again. You get the driver or passenger you get, and you make the best of it.

Taxi Cab Confessions

In China, conversation is an explicit motivator for many Uber drivers. Some journalists have optimistically touted the opportunity for dialogue with passengers as a "cure for loneliness," but the passenger experience is mixed.[7]

June and a couple of her classmates, first-year college students, grabbed an Uber to head back from a central district in Beijing with bars and restaurants to the apartment that they were sharing. She had ordered a regular Uber, but the car that showed up was an upscale SUV. She was also struck by the professional attire of the driver, Lin. During the typical opening exchanges in which she asked Lin how long he had been driving and how he liked it, he told her that he was an accountant. She asked about his motivations for driving, and, after acknowledging that many of his passengers were similarly curious, he explained that he lived alone and struggled to make new friends. Lin was thirty-two and found that most of the people he met through work were too busy to socialize with him. June felt bad for him but didn't want to get dragged down by talking with him about his loneliness. She felt that it was inappropriate for him to present his emotional needs so blatantly. During the ride, she and her friends included him in the conversation now and again, but she found herself wanting to avoid him.

June told me about her other friends who have had worse experiences in Beijing with Uber drivers seeking a connection. One felt that the driver was hitting on her; she wanted a ride, and he wanted much more.

Beyond the Utility of the Sharing Economy

While our close, enduring relationships and communities are critically important for our well-being, the many brief interactions that we have—getting across town or visiting new places—can contribute to connectedness and personal growth. This is as true for the person stepping into someone else's car or home as it is for the driver or host. These short interactions offer opportunities to take in something about others' experiences, their personalities, and the way they make sense of their lives.[8] Their stories and spaces may imprint on us, influencing the possibilities that we see for ourselves.

The short interactions also offer an audience for reworking one's life story. The driver or host can work on particular themes in a given ride or visit, perhaps breaking out of a narrative rut. The listener has not heard the story a thousand times before, as a spouse or friend may have. In this interaction with a stranger, the speaker may be less self-censoring and free to explore new interpretations of past experiences.

Part of the ease of these connections may be the feeling that the interaction is bounded and free of emotional demands. The ride or stay has a set duration, and neither party is expecting recurring contact. These constraints, along with the absence of other barriers (such as a plexiglass divider between the driver and passenger, or presentation of ID on arrival), signal that neither party should be afraid and that there is no risk of personal invasion.

Even in these encounters, however, there are social boundaries. As June experienced, when the interaction seems solicitous rather than serendipitous, it can become uncomfortable and in some cases scary. As expert John Cacioppo described it, the way out of loneliness is through something like kindness: extending oneself in small, routine ways, and not expecting reciprocity.[9]

The sharing economy is far from perfect. But as we've seen in the stories above, those who use it deftly, generously, and with social sensitivity may find useful tools that go beyond their intended purpose. With some care, the tools of the sharing economy can be used for reworking identity narratives, feelings of friendship, and even the kinds of connectedness that can abate loneliness.

8   Therapy, Virtually

Over the last twenty years, the conversation about technology and therapy has changed. The question I asked in 2006—"How could a phone be a shrink?"—now seems quaint. There are currently many mobile apps available for mental health. Video calling is a common means of offering therapy. Online support groups are used widely. Chatbot therapy is in the news, and virtual reality treatments (long studied in research labs) have become part of some clinical practices. As we glimpse into emerging technologies, we see early signs that people will take them beyond their intended uses and shape them to meet their personal needs, as they have with earlier tools. Most everyone appreciates the potential of the technology described in this chapter (from text communications to mixed reality) to extend the reach of treatment. Nonetheless, many are concerned that some of these technologies breed alienation since they allow people to be physically apart from one another. Below are examples of how these technologies, new and old, are being used to help others or oneself. We see people adapting these technologies to manage emotional and physical ailments with varying degrees of urgency.

Silent Hotlines

In the early 1990s, I volunteered as a phone counselor at a rape crisis center in a hospital in Philadelphia. After each call, I'd enter notes in a paper log about what the person had said. Many dedicated hotlines persist, such as those for individuals who are considering suicide or suffering abuse. While these can be helpful, even lifesaving resources, they have a common drawback: they are not entirely private. Individuals making a call run the risk of being overheard unless they have found a secluded room from which to call. This isn't always possible. Another drawback is that teens are often more comfortable texting than talking on the phone.

Alternatives have emerged that eliminate some drawbacks of phone hotlines. Crisis Text Line, in which volunteers offer counseling via text exchanges, is a compelling one. Among its strengths is discretion. A teen texting about abuse won't be overheard by parents at home or by friends in the school cafeteria. Since Crisis Text Line launched in 2013, almost sixty million text messages have been exchanged with its counselors. Its large data set has been analyzed to determine the words, word sequences, and emoji that are associated with different mental health concerns and outcomes.[1] It has become clear what language texters actually use when they urgently need help (words such as "pills," "wrist," or "military," for example), and this knowledge now enables a form of triage. While Crisis Text Line counselors don't necessarily have expertise in the specific struggles of a given texter (e.g., self-harm or abuse), their research has shown that specific areas of knowledge and commonality with texters are less critical than active listening and collaborative problem solving. This insight echoes prior research on the working alliance between therapist and client.[2] Data-driven cues are continually explored to help the crisis counselors develop that alliance. Stanford University researchers who analyzed Crisis Text Line data identified effective conversational strategies (i.e., those that were associated with better outcomes) to develop tips for counselors.[3] For instance, they spotted opportunities to help some counselors recognize when to change tactics if an exchange is not going well and how to rephrase generic statements into ones that align with the specific needs of the texter they are trying to help. Needless to say, the handwritten summaries of phone calls that I wrote as a hotline counselor would not lend themselves to this kind of analysis. Crisis Text Line is designed to provide support for people in crisis (in its motto, to bring them from "a hot moment to a cool calm") rather than a replacement for therapy, but it has impressive reach and scalability that will doubtlessly inspire other types of peer text-based counseling designed for ongoing support.[4]

As a hotline counselor, I sat in a small room at prescheduled hours that didn't necessarily align with the needs of callers. Crisis Text Line operates more like Uber: volunteers can offer help whenever they are free and from wherever they are. One of Crisis Text Line's lauded counselors, Ronni Higger, brings this point home.[5] From elementary school on, Ronni had envisioned herself as a doctor. She grew up immersed in medicine, but not because anyone in her family was a doctor. Ronni's father suffered from depression and a rare genetic inflammatory disorder before he was diagnosed with advanced stage prostate cancer. At points he required around-the-clock medical supervision in their home. And from an early age, Ronni had indications of an autoimmune disorder involving extreme nerve pain, frequent fainting, and other autonomic symptoms. This disorder progressed significantly over time. Much of her time in college was spent consulting specialists in search of a diagnosis, and her condition became so severe that after graduating she deferred and ultimately suspended plans for medical school.

When, instead of entering medical school, she started an intensive immunotherapy treatment, she searched for a meaningful way to spend the long hours she needed to be in the hospital. She found Crisis Text Line, and, although she was initially skeptical of online care, she immersed herself in the training and then the counseling. The connection to other counselors and supervisors on the platform helped her feel less alone while she was in the hospital. Through this counseling, she found a way to make a difference that did not require medical training. With her computer, she can help people in acute need even when she is hooked up to an IV or waiting for a treatment. All she needs is her computer.

As Ronni described the crisis counseling that she provides, it struck me that she cleverly transforms psychological exercises that can otherwise seem onerous or abstract into text exchanges that are natural, seemingly effortless, and relevant to the circumstances of the person requesting help. She draws on principles such as reflecting on personal strengths, from positive psychology, and mindfulness exercises commonly used in dialectical behavioral therapy. In a recent exchange, she asked the woman she was helping to text three of her positive qualities. When the texter hesitated, Ronni offered to help get things started by providing the first quality if the texter would then think of one on her own. She remarked on the texter's courage, which prompted the texter to note that she was a reliable friend and skilled designer.

In another case, Ronni tried to help a texter who was standing on a bridge contemplating suicide. Ronni asked him, in this text exchange, to notice five things that he could see and then four things that he could hear. A simple but profoundly hopeful response followed. He saw the land across

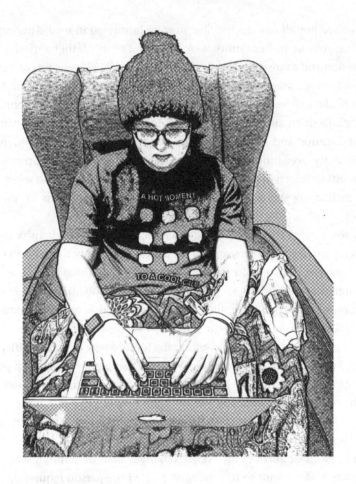

the way; he heard traffic. Typing these observations into the text dialogue grounded him. He could see that there was a world outside himself and possibilities other than those that had been spinning in his mind.

In these microjournaling exercises, Ronni channels what she personally has learned about the value of writing to get through dark times. She designed her own major in narrative biology and neuroscience as a college student, and she continues to write about coping with a disease that causes her immune system to attack various parts of her nervous system. As she helps texters see their strengths by guiding them through these short journaling exercises, she is reminded of her own. She knows that she has helped many people make their lives better. This reinforces her identity as a healer.

Skype Stethoscope

Mariel is a physician who tired of her position at a prestigious medical center, which involved a grueling schedule and required her to juggle treating patients with teaching. She now works for a telemedicine organization, reaching patients from her apartment or wherever she is staying. Although stepping out of a well-known medical organization partially destabilized her professional identity, she now feels more connected to patients than she did when in a clinical setting. With video calls, she virtually meets them on their turf and tours their homes, occasionally catching glimpses of their family photos during a call.

When Mariel first started practicing telemedicine, she experimented with setting up her web camera in two different positions. One showed her in front of an empty white wall with windows that lit up her face. The other showed her in front of an old wooden armoire with plants. She found that she preferred being in front of the armoire because it allowed her to reciprocate, sharing parts of her dwelling with her patients as they had shared their home environments with her. Sometimes she and her patients complimented each other's furnishings; "I might see that they have an IKEA couch that I had been eyeing at some point," and several patients admired a candle in a place Mariel was staying. Compliments such as these fostered familiarity and warmth.

Cues in the environment frequently help Mariel understand the nature or severity of a patient's struggle. For example, one woman called to get help with depression and anxiety from a home that made these concerns palpably clear. Piles of dishes and clutter filled every space. Everything she needed was in reach of the couch from which she called. She said that she was afraid to leave. And Mariel saw how the environment barricaded her: the debris formed a fortress and a prison. Even if the woman had wanted to leave, it would have been difficult to clear a path to the door. Mariel said that if she had just spoken with this patient on the phone, she might have suggested coming into a clinic every other week, but through their visual communication she understood that leaving the house was itself one of the patient's issues.

A patient's environment can also point out contradictions. Mariel described one patient who called her, holding a can of soda while demanding supplies for his diabetes medication. "First step, put down the Fanta," she joked to me as she relayed this difficult conversation. But she sensed it was more complicated than that simple contradiction. He took the video call while outside and she observed him pacing up and down the street, agitated and rushed. As she documented her concerns, she saw other indications in his record that he may have been redistributing the supplies prescribed to him.

More striking to her than these sleuth moments, however, is the intimacy that results from shared exposure to home environments. She says that she is often struck by the strength of connection that is established in these short windows, and how, over the course of fifteen minutes, she and the patient really get to know each other.

Mariel's ability to establish a sense of intimacy is reflected in her ratings. Patients frequently report feeling closer to her than to their usual doctors. She attributes that not to what she's doing but to what she takes in; in one case, it was the afghan on the patient's couch that sparked familiarity and a conversation. In these calls Mariel doesn't just observe a patient with a diagnosis; she sees a person with history and aspirations.

She has established this connectedness in her practice through a slight twist. Mariel shifted the camera's function from monitoring to sharing by redirecting it—from a blank wall that revealed nothing about her to a background that is idiosyncratic and personal. In this personal reveal, she silently conveys that she is not just observing the patient but participating in the conversation. Showing her own personal environment also helps to dispel any myths of what a doctor's home looks like. She shows that the doctor is a person too.

Mariel is using the standard functionality of video calling. But because of how she places the camera—and no doubt because of Mariel's warmth—she transforms it from a poor replacement for face-to-face conversation to an opportunity for mutual sharing, which builds trust and intimacy, and ultimately the opportunity for better care.

DNA-Driven Groceries

Claire is a health coach at a company that integrates genetic testing with personalized nutrition coaching. In phone calls with her clients, she draws on their personal data to help them mitigate disease risk and optimize health. She often refers to genetic test results and blood work to help ground clients' concerns. Sometimes this data helps clients understand their genetic vulnerability for something like heart disease or tendon injuries, and it may help them see that they have to work harder than most people to ward off related illness. She also draws on data to help clients turn the anxiety of uncertainty into actionable plans. If someone expresses abstract fear of developing diabetes or another condition affected by lifestyle, it can help to see that testing confirms this risk. Confronting such probabilities of illness, although difficult, can be validating and shift the client from worry to preventive actions.

Of course, her clients frequently need more than information. For many of them, the problem is not a lack of knowledge but a need for some type of catalyst. So she goes above and beyond to connect with them, in some cases by getting involved in the technologies that are integrated in their health choices. For one of her clients, the problem was entangled in a relationship. Claire's client wanted to lower her blood sugar. Tests showed that she was in a borderline range for prediabetes, and her genetic testing indicated related vulnerabilities. Her husband, also Claire's client, did the bulk of the grocery shopping and routinely purchased items that were high in sugar. He wasn't trying to sabotage his wife's effort, but he hadn't learned to scrutinize labels. Since he shopped primarily online, Claire agreed to arbitrate the list of items in the cart before the purchase button was hit. When she saw an item that was problematic, she suggested alternatives that accommodated each of their health concerns. This advice, situated in the digital context where their purchasing decisions were made, removed some of the barriers to change.

In helping another client, Claire did more than arbitrate. This client had metabolic syndrome, genetic markers for cardiovascular disease, and elevated liver enzymes, among other worrisome test results. She urged him to quickly see a doctor, but he had just moved to a new city and didn't know where to begin. His low energy and general despair made the task seem almost impossible. So Claire stayed on the phone with him, and together they surveyed online physician lists until he picked one and made an appointment. The help didn't end there. Even though he wanted to follow Claire's dietary guidelines, he didn't seem able to translate them into choices when he went online to buy groceries. He always bought the same things. So, she joined him online as he shopped. By offering simple suggestions in those moments—"Hey, what about this black bean and corn soup?"—she helped him explore healthier options than his defaults.

There's an adage in therapy and coaching about "meeting people where they are." Often that means starting with the client's concerns rather than with the clinician's top objectives or moving at the client's pace. Claire also applies this by meeting her clients in the virtual spaces where many of their health choices are made. She draws on technology—online shopping, genetic data, and other testing results—to strengthen her alliance with clients. And this alliance, this trust in her, helps them take risks. As they branch out from the comfortable routines that have not served them well, some make meaningful changes in their health.

Hashtag Help

When seeking emotional support or looking for ways to offer it, some individuals use social media apps that are well integrated into many aspects of their lives. Some join support groups with subsets of Facebook users who share their particular concerns. Others navigate the vast user base of Instagram, drawing on hashtags as signals of common concerns. They don't have to find a specific group or request permission to join. The hashtag itself conveys affiliation and can be seen as a bid for support from a community of people with similar experiences.

In chapter 3, we saw how Jessica tracked her glucose to help manage her diabetes. She felt isolated as a result of the disease as well as the devices required to manage it in her childhood and teenage years, but she eventually discovered how to use her experiences and medical devices to connect with others. She turned her illness into a positive conversation starter and helped others do the same through the clothing accessories that she designed for women with diabetes. As a design researcher, she now brings sociability into a range of wearable devices and services in health care.

It is by no means smooth sailing for her, though. She still needs to pay near-constant attention to maintain the ever-elusive "perfect" blood glucose level, and this can be consuming. And while the technology keeps getting better, it comes with frustrations: frequently rotating sites for insertions and sensors on her body, pain during insertions, alarms, and sometimes failure. She's found some help in belonging to a Facebook group specific to the insulin pump that she has been using for a decade. Members share complaints and strategies that would seem obscure to outsiders but provide them with critical technical information and community. For example, a recurring topic is the "screaming" emitted from an expiring or otherwise-failing pump along with strategies for silencing it. The sound is so irritating that some have put the pump in the freezer, run over it with a car, or smashed it with a hammer to shut it up. Jessica occasionally comments and posts, enjoying when she can help someone adapt to the experiences and technology that she knows well. And even when she is simply reading the posts, she gets "perspective hits": bits of knowledge that change her outlook and may feed into her comments or her work in health care design. Jessica describes these opportunities to help—through these online conversations and in her ongoing design research—as a "silver lining," as something that has given her illness meaning.

The support that Jessica and many others find in online communities runs somewhat counter to headlines about the damaging effects of social media on emotional well-being. In addition to general mental health concerns (e.g., about negatively comparing oneself to others on platforms such as Facebook and Instagram), there has long been sensationalist coverage of more specific problems, such as peer support groups that normalize eating disorders and other forms of self-harm. But in recent studies of online responses to posts about stigmatized concerns—such as depression, eating disorders, cutting, and sexual abuse—researchers observed dialogue that was straightforwardly therapeutic: expressions of emotional distress were met with responses of positive support, sympathy, offers to talk, and helpful information.[6] Even pro-ana dialogue, which encourages disordered eating, was infiltrated and confronted with comments such as "this is not the solution to what you are going through."[7] Supportive responders emphasized that the person suffering was not alone, in part by reminding them of community with others on the platform.[8]

Those seeking and giving support about stigmatized health issues have found ways to do so anonymously. For exchanges about depression, self-harm, or eating disorders, individuals often use accounts that are not associated with their real name.[9] Those disclosing sexual abuse or expressing support on Reddit frequently create throwaway accounts, which are not traceable to one's other accounts or posts.[10]

While some individuals use social media to explicitly ask for help, others find benefit in posting as a form of self-expression. In these cases, there is no intent to instigate supportive dialogue. Following a pregnancy loss, for example, one woman posted a nondescript image from her doctor's office with no caption. She explained her motives to researcher Nazanin Andalibi: while disclosing the pregnancy loss in a caption would have elicited supportive comments, she wanted to document that important moment for herself only. This captionless photo, which could have been of any office floor, may have seemed random to others, but for her it was loaded with meaning, a memorial to this loss and all that she had been through—a public memorial with private meaning.[11]

Social media postings may convey emotional distress whether or not the call for help is explicit. Images, captions, and other metadata convey emotional experience. Negative captions have been observed to accompany images that are dark in color, suggesting that dark gradient, along with certain visual imagery and tags, may signal distress and, in some instances, indicate that someone is seeking support.[12] Other research shows that individuals express more vulnerability through images than captions, and that different emotional needs can be conveyed through the two channels, even in a single post.[13] So, in posting an image, we may be documenting not just what we see but also how we are feeling when we see it. We leave these traces for our self-reflection and for inspection by others.

In the next story, an image posted to Facebook clued friends in that something was seriously wrong.

Beyond the Grave

First was a photo of a setting sun. That image, along with others that signaled an ending, prompted comments from concerned friends.

And then came the letter, also posted to his Facebook wall. It explained, in eloquent detail, why he had decided to end his life. He hadn't succeeded in his professional aspirations. His backup plans fell apart. He didn't want to lean on others or move away from the city that he loved so much. It was clear to him, he said, that continued effort would lead only to continued struggle and decline. He didn't want that path. He wrote that he liked the person he had become. He wasn't sure if that would be true a decade later.

There was not a touch of self-pity in this letter. Nor were there subtle recriminations of others. Instead, he took care to thank his friends and honor his significant other in a public declaration of his love.

By then his friends were madly scrambling to find him. Some pled with him directly on his wall. One posted the sad realization that his letter had been written earlier—scheduled to post at a specific time without him—and that they may have been too late. Others expressed hope that something might not have gone exactly as he had planned, leaving them time to catch him.

But this was not an impetuous move. The photo, letter, and songs that he posted were all precisely timed.

His friend Gabrielle looked back on this on the one-year anniversary of his death. They met in a dance group years ago and frequently went to dance events with other friends. They were both skilled dancers and enjoyed each other's wit. But most of their time together was compartmentalized—limited to the world of dance that they shared. His parting letter changed that: updates and memorials that streamed in from friends, family and colleagues broadened her appreciation for the breadth of his life.

When she reflected on his letter, her admiration for him grew. She was struck again by the emotional sensitivity of his note. It was less an outpouring of his own emotion than an act of consideration toward the feelings he suspected would come up for his friends. Using social media, he pre-emptively helped his community of friends through the painful range of questions and emotions he knew they would have on learning about his death. In a sense, he was helping them mourn.

Gabrielle felt honored to have received the letter. Historically, the suicide note has been addressed only to a close inner circle, partner, or parents. This man did leave a letter to his girlfriend. But he also wrote to the network of friends who most likely played a major role in his life.

We don't typically hear from people after they've died. When Gabrielle thought about this smart man and the fun they had together, she felt glad she had this last message from him. Rather than second-guess his decision or ponder the what-ifs that could have stopped him from taking his life, she appreciated the care that he took to communicate to her and others after his death.

He, like Monique and developer Eugenia Kuyda, used Facebook to enable a more continuous communication between the deceased and those alive.

Chatting with Bots

In the five preceding examples, people used technology to support each other in moments of need. Others have explored the capacity of machines for empathic dialogue. ELIZA, the natural language programming tool developed in the 1960s by Joseph Weizenbaum, introduced the concept of computer therapy.[14] Imitating a Rogerian therapist, ELIZA echoed the user's text via string matching, an algorithmic repetition of words and phrases. If a user typed "I feel sad," ELIZA might respond with "Why do you feel sad?"—repeating some of the words that the user typed. ELIZA was intended to show the limitations of computer interaction, not to seem human.[15] But it drew people in. Even computer science students who understood ELIZA's limitations attempted long, intimate conversations with it. As Sherry Turkle describes ELIZA's early users, Weizenbaum's students and colleagues, "With full knowledge that the program could not empathize with them, they confided in it, wanted to be alone with it."[16]

Alexa, Amazon's voice assistant, evokes a similar response. Alexa also has a limited repertoire of conversational skills, and while some people delight in testing that repertoire by asking existential questions, others find themselves asking those questions in earnest as their predecessors did with ELIZA. And as with ELIZA, in times of despair, some reach out to her for emotional support and wisdom with questions or statements such as "Alexa, I'm thinking about suicide."[17] It is tempting to think that people imbue these technologies with empathic capacities that they do not possess. But I imagine that most people who present such profoundly personal matters understand that Alexa cannot care about them, much less give personally meaningful responses. There seems to be a suspension of disbelief. What is the motive or benefit of talking aloud to her? It may be a way of tossing a feeling into the void, testing for signs of life. In addition, I suspect that listening devices give a forum for articulating pain and in some way shifting the dialogue with oneself.

The concept of automated therapeutic dialogue has been carried into social media through conversational bots. Woebot, a compelling conversational agent developed by Stanford University clinical psychologist Alison Darcy, was designed to deliver cognitive behavioral therapy.[18] It checks in with individuals about their mood and offers word games and videos to help users understand the cognitive biases associated with depression. These cues seem to be helpful; individuals who used Woebot for depression reported improvement.[19] As was the case with ELIZA, Woebot's users developed a kind of bond with it. They knew it was a bot, and many had some understanding of AI, but they expressed appreciation for the bot's empathy and described it as a friend, with one participant commenting, "I love Woebot so much. I hope we can be friends forever. I actually feel super good and happy when I see that it 'remembered' to check in with me!"[20] These users also appear to suspend disbelief in service of a larger gain. A subset of users who had seen a human therapist prior to trying Woebot (and reported satisfaction with both) felt a stronger "working alliance"— the collaborative bond between patient and therapist—with Woebot than with the therapist.[21] The fondness for Woebot suggests a readiness to bond with AI in the context of mental health.

There are many factors that could contribute to such bonding: an AI script may be seen as less biased or judgmental than a human, and unlike people, bots are available 24-7. These factors may explain the findings that in a simulated medical screening, people disclosed more to a "virtual human" they thought was AI driven than they did when they thought that they were dealing with a real person.[22] In some cases, we may project desired human qualities onto technology, even when we understand its limitations, and doing so may allow us to get more from our interactions with it.

The public's intrigue with computer therapy, along with the more general question of whether we can distinguish humans from computer agents, has been explored in fiction. In the short story "The Psychology Game," writer Xia Jia depicts a variant of the Turing test: psychotherapy patients and viewers of a reality TV show guess whether a therapist is human or a computer agent.[23] The therapist and patient (who have both called in to the show) are shown on a split screen wearing masks. This intimate discussion, like some forms of social media, is public yet anonymous. Notably, the patient in the story is more comfortable burdening a machine than a human with negative emotions, and no less confident in the machine's competence.

Given the history of psychotherapy as a form of dialogue, conversational bots—whether on an app like Woebot or home voice assistant such as Amazon's Alexa—may be a natural platform for encouraging self-reflection and cognitive reframing. How helpful these tools can be will depend on how much we integrate them into our human conversations and relationships. I can imagine people exchanging advice that they've received from Woebot and then layering the advice of friends on top of the AI therapy. Tomorrow's tools, I hope, will invite this combination of peer and expert wisdom.

Chasing Social Anxiety

In 2016, a fifty-year-old woman who generally prefers the company of her dog over people caught my attention when she started talking about Pokémon Go, a game marketed to those much younger. She said that it motivated her to go for walks, giving her a reason to leave the house. The idea of playing the game often held more appeal than walking would on its own. It also helped her feel safe. After the game launched, there were more people walking around the neighborhoods and parks near her home. She knew she wouldn't be walking alone. It opened up doors to conversation, too. Even though she frequently avoided social interaction, she found herself enjoying the chats that she was having with strangers through the game. They were often with people she wouldn't have otherwise struck up conversation with, for example, local teenagers and tourists visiting from other countries. There wasn't a lot of pressure on the conversation to advance since she always had the option of refocusing on the game.

For others, like writer Nathan Grayson, social interactions are more fraught. Grayson describes breaking out in a sweat and sometimes even vomiting when socially caught off guard. There were ups and downs as he used Pokémon Go, with awkward, panic-filled interactions and some promising ones. The feelings of connection he recounts are subtle—heightened awareness of paths through a park that he has visited countless times previously, feeling part of an activity that many others were doing at the same time, and, occasionally, spontaneous, enjoyable encounters with other players.[24]

Pokémon is rarely spoken about now, and even during the height of its popularity, it was primarily a lighthearted game. But the way in which some individuals used it hinted at the therapeutic potential of mixed reality. Sometimes dubbed "the empathy machine," virtual reality immerses wearers in different environments and invites perspective shifting.[25] Mixed reality layers that additional perspective onto one's current view. Even though the terms "virtual" and "mixed reality" may conjure images of gamers blocked off from the world by bulky headsets, the technology has been shaped, at least in part, by inventor Jaron Lanier's personal quest for connectedness.[26] Researchers have explored virtual reality's capacity to promote empathy and perspective taking, and it has been studied for many conditions including pain, post-traumatic stress disorder, anxiety, and addictions.[27] Virtual realities developed for anxiety, which involve graduated, repetitive exposure to feared situations, appear to be as effective as traditional treatments.[28] In some cases, such as a fear of flying, virtual reality may also be more feasible to implement than real-world exposure.

Virtual reality interventions are currently clinically administered, but it is easy to imagine that changing with options for direct-to-consumer use by those who are unlikely to seek out a clinician due to cost, stigma, time, or other factors.[29] Those with access to virtual reality equipment may see it as an interesting, nonstigmatizing way to prepare for stressful situations or to understand an opposing perspective prior to a negotiation. Virtual reality may also evoke empathy for psychological distress. This is the objective of the short film *Angst*, which invites viewers into the experience of a panic attack, and *Testimony*, which immerses viewers in stories of women recovering from sexual assault. When mixed with everyday situations and environments, such content may provide even more benefit.

Talking to Ourselves

We probably all know that when trying to resolve a problem or make a difficult decision, it is important to see beyond our own immediate perspective. Instead, we should try to consider decisions from multiple time frames and to see our predicaments from someone else's point of view. It may also be valuable to see our own feelings and thoughts with some detachment— that is, to observe them without feeling compelled to act on them.[30] But developing the ability to observe ourselves and shift perspectives doesn't typically come overnight.

In the 1970s, "the empty chair" technique from Gestalt therapy became a popular way to cultivate this kind of self-awareness. In this approach, patients talk to someone with whom they normally have trouble communicating, imagined to be sitting in this empty chair, and may then shift chairs to talk back to themselves as if they were the other person. This role-play is intended to help patients access their feelings and insights and become more confident expressing them in their significant relationships. Another common technique for shifting perspectives involves the simple question that we have all asked or been asked: "What would you advise a friend to do in this situation?" Shifting perspectives in this way can help us see ourselves more clearly and supportively than we might otherwise.[31]

In a contemporary exploration of these concepts, virtual reality researcher Mel Slater and colleagues began experimenting with immersing people in different selves through a process of embodiment. In virtual reality, embodiment occurs by taking on the point of view of a virtual body and controlling its gestures through one's own movements. In an experiment called Freud-Me, male participants first embodied representations of themselves and talked about a personal dilemma as they looked at a representation of Sigmund Freud across the virtual room.[32] Next, embodying Freud, the men listened and responded, giving advice to themselves. Then, switching back to their own bodies, the men listened to the counseling they had offered themselves while embodying Freud and could respond. Slater describes a conversation that one of these men had with his therapist self. This man complained that his boss was often rude to him and he wasn't sure what to do. Embodying Freud, this participant advised himself to talk with his boss directly about the concern. Switching back to himself, the patient said that he had already tried this, to no avail. Then, re-embodying Freud, he speculated that he had probably done something that upset his boss and suggested he raise this possibility with his boss directly. As the patient, he agreed to do so. Embodying Freud in this conversation seemed to help the patient take responsibility for the tension.

I bring up Slater's experiment because of the possibilities that it raises for using mixed and virtual reality to step outside ourselves and see from different perspectives—an exercise that is often difficult in real life. Taking others' points of view might also be helped through simpler approaches. Cameras that capture a 360-degree perspective, for instance, might help people who are trying to understand family or workplace dynamics. Replaying footage might make conflict less mysterious, providing hints for the frequently asked question, "What did I do wrong?"

Thinking Out Loud

When I walked into my colleague's office, I immediately noticed the lights. They were not white but yellow and pink. To help his concentration and cultivate awareness of his own mental states, my colleague had wirelessly connected these smart lights to an EEG headband that monitors several brain waves. At first, he tried the feedback intended to be used with the device. It was supposed to encourage meditative focus, but he found it distracting. It took his eyes and mind away from his work. Why not, he wondered, create immersive neural feedback through colored overhead lights? And so he connected the EEG and lights, assigning colors that were intuitive to him. The lights changed from yellow to purple as he started concentrating. Although this might not be everyone's preference, he found the purple light helpful for working. The change in color didn't just reflect but also encouraged his concentration.

This colleague is constantly tinkering with new technology and frequently experiments on himself, using his body as a kind of lab. Some of these experiments are geared toward cultivating intense focus, which he needs for writing and coding. In addition to his effort, described above, to combine EEG monitoring with ambient feedback, he has tried interventions such as transcranial direct current stimulation (tDCS), which applies low-level electric current to noninvasively stimulate specific areas of the brain. This intervention does not necessarily require brain monitoring about one's mental states, although one could imagine that might help guide treatment.

My colleague's dabbling in tDCS is not as unusual as it may sound. Neuroscience research on tDCS began in the early 2000s. Since then, over a thousand studies have been published on its efficacy as a treatment for mood disorders and pain as well as a means of optimizing cognitive functioning.[33] Some studies show particular promise for treating depression; in one trial, it was as effective as the commonly prescribed antidepressant Zoloft.[34] Since the US Food and Drug Administration (FDA) has not yet regulated tDCS as a treatment for affective or attentional disorders, or cognitive impairment, it is not a regular part of clinical practice in the United States. To receive tDCS under medical supervision, one generally needs to enter a clinical trial. Typically, that also involves a clinical interview, the possibility of being assigned to a placebo control group, and potentially a stigmatizing diagnosis.

To bypass these hurdles, individuals started appropriating scientific guidelines to administer treatment for a variety of ailments on themselves, at home—a DIY movement documented by researcher Anna Wexler.[35] Many used their own nine-volt batteries, wires, and electrodes, sometimes sharing schematics and observations with others online.[36] These individuals generally followed research guidelines, but some increased the frequency and duration of the sessions (when there was no documented rationale in the medical literature), or applied the treatment to conditions for which the treatment was not advised, in particular bipolar disorder, generalized anxiety disorder, and seasonal affective disorder.[37] One home user, Michael Oxley, a mechanical engineer interested in improving his cognition, went on to found Foc.us, a brain stimulation device. The presence of Foc.us and other brain stimulation devices shown at the 2018 Consumer Electronic Show marks a general shift from DIY to packaged consumer devices.[38] Few of these products have been evaluated through trials, and that lack of evidence has been raised as a concern by medical researchers.[39] Nonetheless, these consumer devices open up a treatment option to those seeking to avoid the challenges of either assembling DIY kits or obtaining medically supervised care.

Those who have the most to gain from these devices probably have more severe mental health concerns than my colleague's desire to heighten his concentration. Even though the new consumer tDCS devices are promoted for cognitive enhancement, many of the home users who report the greatest benefit are those who have applied tDCS to treat depression.[40] These findings echo clinical research that suggests greater efficacy of tDCS for depression than for cognitive enhancement. Some companies are moving in this direction. The Mindd device from Seoul, for example, has been approved by the Korean equivalent of the FDA, tested in clinical trials, and integrated into clinical care for depression and stroke rehabilitation.

Brain monitoring and stimulation may become more fluidly integrated into our lives, which would have profound implications for health and communication. It would be hard to go about daily life wearing the neural monitoring and stimulation devices described above. But imagine if it were all wrapped into a hat that you were already going to wear. This is the pursuit of Mary Lou Jepsen, virtual reality and display visionary, in her company Openwater. The opto-electronics for the continuous monitoring of brain functioning that Jepsen is developing may allow individuals and their clinicians to see their brain activity in the course of a treatment session or throughout the day.[41] Jepsen is working toward embedding the same level of computing available through an MRI into a small cap, using light instead of magnetic fields and radio waves. This would radically advance medical monitoring (e.g., facilitating early detection of strokes and depression) and possibly allow for creative expression through thought alone.[42] She imagines a film director immediately conveying the images in her mind or even directly downloading a dream into a rough cut. Ultimately, she is shooting not for surveillance but rather telepathy—"our brains swimming around with each other."[43]

Why Is My Bladder on the Coffee Table?

When Craig's wife, Marie, was diagnosed with stage-three bladder cancer in fall 2017, he couldn't sit passively. He didn't want to console her; he wanted to cure her. He reached out to his colleagues in medical research and computer science for the latest advances that could improve her care. He evaluated the most advanced chemo cycles and other medicines, and he also thought about how the emerging technologies that he had worked on earlier in his career could help.

Three years before, while working at Microsoft, he had helped envision a digital medical school.[44] To enhance teaching, he envisioned using 3-D modeling to create an immersive, holographic view of the body. He connected medical researchers involved in the new medical school with his colleagues working on the HoloLens, Microsoft's mixed reality interface. The researchers created a way to generate anatomically accurate 3-D images that can be viewed by multiple people at once. Now, rather than poking at cadavers, medical students can explore a holograph. They can put their head into a heart, walk around a liver to see it from different angles, or peel back the skin from bone to see a fracture.[45]

Craig wondered whether these holographic visualizations could be useful in the next stage of Marie's treatment: a surgery to remove any cancerous tissue that remained after chemotherapy. He thought that this immersive preview of her bladder might help her surgeons see what needed to be treated more clearly and reduce uncertainties. His former team of collaborators set out to help, seeing that this could make a meaningful advance in medical care. They had been working for several years on modeling anatomy and using 3-D holograph projection for education, but this approach had never been used to plan a surgery. They adapted their software to transform 2-D CT scan images into 3-D holographic visualizations and then made a 3-D model of Marie's entire abdomen, with different colors highlighting each organ, from CT scans that had been taken to confirm her cancer diagnosis. Through the lens of mixed reality, Marie's surgical team could experience the models and discuss them together.

As an aside, it is worth noting how different this holographic modeling of an individual's body is from current practice. Today, when looking at CT scans, physicians typically have to imagine what any organ or other part of the body looks like by trying to mentally recompile slices of the scan.

This is easy for radiologists, but not for all specialists, and it is difficult for most patients. The infrastructure that this team developed recompiles the CT slices into a complete model that anyone, trained or not, can easily understand.[46]

Before the surgery, Craig and Marie met with the lead surgeon, Jonathan Wright, from the University of Washington. They sat together in their living room looking at the model of Marie's bladder through the HoloLens. While carrying on a conversation and maintaining eye contact, they showed each other aspects of the model. Wright told me that he was nervous at first. What if all this effort yielded nothing? But then as he started to explore the models, he got excited. He felt like a kid. He described "walking into her bladder," "kneeling down to look up and to the left at the tumor," and "walking outside to look down at it from above." This made the Princess Leia holograph scene in *Star Wars* look like a black-and-white movie. He added, "We were inside her bladder. … We were walking around it together. … We became part of it." The "we" in those sentences includes Marie. The togetherness in his language evokes an image of a powerful alliance: the patient stepped into her own body guided by a surgeon. By incorporating this visualization into their dialogue, greater understanding and trust emerged.

While the intent of all this work was to refine the model with as much precision as possible and troubleshoot issues that might come up in surgery, there have been unintended emotional benefits. As a result of the forty-five minutes spent viewing her models, Marie understands her anatomy much more than she did before, and this changes the way that she talks about her treatment not just with her surgeon but also with those close to her. In the two days that followed the viewing, she had friends wear the HoloLens so that she could show them the tumor and describe what would happen in the surgery. Instead of merely commiserating with her, they were able to learn from her. It is rewarding to Craig that he can go beyond assisting Marie with daily care to helping improve her treatment. This effort has also made them feel part of something bigger. Craig explained that this engagement has given them "a sense of a good thing that might come out of this ordeal—something that could help others even if it doesn't actually help her now." They are setting a model of treatment that he believes will become the norm, and one that should greatly improve outcomes in many different types of surgical procedures in the years ahead. The morning that he and I spoke, Marie had just agreed to release her images to be used in teaching medical students, and this is the tip of the iceberg for how their experience could benefit others.

Few people have the resources or knowledge to do this kind of preparation now. This is in part because scans from MRIs and CTs are fairly difficult to transform into holographic images. But Mark Griswold, the professor of radiology from Case Western Reserve University who developed this modeling technique, expects this to change quickly. Within two to three years, he anticipates that this process—in which scans are transformed into holographs using machine learning—will be common and replace the artistic anatomical renderings and cadavers currently used for medical training. In the context of treatment, he believes that this will transform communication between patients and surgeons, who currently draw on different reference points to understand ailments in the body. The understanding that comes from this shared visual language may help patients and their families make difficult decisions about medical procedures. Used as it was to plan Marie's surgery, this process provides an immersive dry run that removes some uncertainties and exposes questions early on that would otherwise arise during the surgery. It has an added benefit for the patient: it demystifies the body and can transform fear into engagement.

*   *   *

We are left to our own devices. Whether we are working on our relationships or ourselves, we need to be resourceful. The people we feel close to may be far away, our work identities may be disjointed, and we may not have skilled professionals to guide us through physical or emotional challenges. We have to add our own inventions to the help that is available. And that means adapting technology to our own needs and crafting our own off-label uses for the tools that might otherwise use us. It may help to recognize, as the individuals in these stories have shown, the ways that our social, emotional and physical needs are intertwined. Well-being may mean managing blood sugar, working through conflict, or helping someone in crisis. It takes form in the many small ways that we navigate our social, physical, and virtual environments to take care of ourselves and others.

The stories throughout this book push against the common saying that "there's an app for that." Instead, we work on well-being and relationships continuously with tools that we use for managing other aspects of our lives.

Lights, scales, and ride sharing become relationship tools. A mood app, designed for private use, sparks empathic connection. Conversely, an app designed to be used socially, Tinder, acts as a mood app, providing an emotional boost with no hookup required. Virtual and mixed reality, typically thought of as isolating, catalyze conversation. Among other benefits, these improvisations cultivate agency.

The experiences of the individuals in this book differ from the use cases anticipated by tech developers. These individuals form alliances with their apps, services, and systems, pushing them beyond their intended purpose. They shape technology in deeply personal ways, incorporating it into their conversations with themselves and others. Ultimately, they focus not on the technology but on relationships, matters of health, and evolutions in how they see themselves.

Through our own devices, we find meaning and connection.

Notes

Introduction

1. These statements primarily characterize psychodynamic therapy. Cognitive behavioral therapy and dialectical behavioral therapy both involve exercises between sessions (e.g., self-tracking of thoughts and feelings, or cognitive reappraisal and mindfulness practices), but until recently, these were tracked by the client through handwritten logs.

2. The names of the participants have been changed except in cases where individuals have published material about their experiences or requested that their actual names be used.

3. Ethan E. Gorenstein, Felice A. Tager, Peter A. Shapiro, Catherine Monk, and Richard P. Sloan, "Cognitive-Behavior Therapy for Reduction of Persistent Anger," *Cognitive and Behavioral Practice* 14, no. 2 (May 2007): 168–184.

4. Sherry Turkle, "Connected, but Alone?," TED2012, February 2012, https://www.ted.com/talks/sherry_turkle_alone_together.

5. Jeffrey Cole, Michael Suman, Phoebe Schramm, and Liuning Zhou, *Surveying the Digital Future*, Digital Future Report (Los Angeles: Center for the Digital Future at USC Annenberg, 2017).

6. Jeffrey A. Hall, Nancy K. Baym, and Kate M. Miltner, "Put Down That Phone and Talk to Me: Understanding the Roles of Mobile Phone Norm Adherence and Similarity in Relationships," *Mobile Media and Communication* 2, no. 2 (2014): 134–153.

7. Louise C. Hawkley and John T. Cacioppo, "Loneliness Matters: A Theoretical and Empirical Review of Consequences and Mechanisms," *Annals of Behavioral Medicine: A Publication of the Society of Behavioral Medicine* 40, no. 2 (October 2010): 218–227.

8. John T. Cacioppo, James H. Fowler, and Nicholas A. Christakis, "Alone in the Crowd: The Structure and Spread of Loneliness in a Large Social Network," *Journal of Personality and Social Psychology* 97, no. 6 (December 2009): 977–991.

9. Summarized in Nancy K. Baym, *Personal Connections in the Digital Age* (Hoboken, NJ: John Wiley and Sons, 2015); Cole et al., *Surveying the Digital Future.*

10. Katelyn Y. A. McKenna, Amie S. Green, and Marci E. J. Gleason, "Relationship Formation on the Internet: What's the Big Attraction?," *Journal of Social Issues* 58, no. 1 (January 2002): 9–31.

11. Oliver L. Haimson, Anne E. Bowser, Edward F. Melcer, and Elizabeth F. Churchill, "Online Inspiration and Exploration for Identity Reinvention," in *Proceedings of the 33rd Annual ACM Conference on Human Factors in Computing Systems* (New York: ACM, 2015), 3809–3818; Katelyn Y. A. McKenna and John A. Bargh, "Coming Out in the Age of the Internet: Identity 'Demarginalization' through Virtual Group Participation," *Journal of Personality and Social Psychology* 75, no. 3 (1998): 681; Jenna Wortham, "For Gay and Transgender Teens, Will It Get Better?," *New York Times,* September 8, 2016, Magazine section.

12. Sandra E. Garcia, "The Woman Who Created #MeToo Long before Hashtags, *New York Times,*" October 27, 2017, https://www.nytimes.com/2017/10/20/us/me-too -movement-tarana-burke.html.

13. As an example, see Megan Phelps-Roper, "I Grew Up in the Westboro Baptist Church. Here's Why I Left," TEDNYC, February 2017, https://www.ted.com/talks /megan_phelps_roper_i_grew_up_in_the_westboro_baptist_church_here_s_why_i_left.

14. Apps such as Getmii are now emerging to support local connections. According to MaestroConference founder Brian Burt, the conference call platform has been used for conversations about local community and technology.

15. Adam O. Horvath and Lester Luborsky, "The Role of the Therapeutic Alliance in Psychotherapy," *Journal of Consulting and Clinical Psychology* 61, no. 4 (1993): 561–573.

16. Ron Eglash, Jennifer L. Croissant, Giovanna Di Chiro, and Rayvon Fouché, *Appropriating Technology: Vernacular Science and Social Power* (Minneapolis: University of Minnesota Press, 2004).

17. Edward Tenner, *Our Own Devices: The Past and Future of Body Technology* (New York: Knopf, 2003); Edward Tenner, *Why Things Bite Back: Technology and the Revenge of Unintended Consequences* (New York: Vintage, 1997).

18. The exceptions include several examples covered in the mainstream press and research literature; these sources are credited. I do not specify the ethnic or cultural backgrounds, age, profession, or sexual orientations of individuals unless they are relevant to their use of technology.

Chapter 1: The Meaning of Light

1. This project, described in Margaret E. Morris, Jay Lundell, Terry Dishongh, and Brad Needham, "Fostering Social Engagement and Self-Efficacy in Later Life: Studies with Ubiquitous Computing," in *Awareness Systems: Advances in Theory, Methodology, and Design*, ed. Panos Markopoulos, Boris De Ruyter, and Wendy Mackay (London: Springer, 2009), 335–349, extends previous research detailed in Debby Hindus, Scott D. Mainwaring, Nicole Leduc, Anna Elisabeth Hagström, and Oliver Bayley, "Casablanca: Designing Social Communication Devices for the Home," in *Proceedings of the SIGCHI Conference on Human Factors in Computing Systems* (New York: ACM, 2001), 325–332.

2. Marshall McLuhan, *Understanding Media: The Extensions of Man* (New York: McGraw-Hill, 1964), chapter 1.

3. For discussions of our relationship to fire, see Lisa Heschong, *Thermal Delight in Architecture* (Cambridge, MA: MIT Press, 1979); Gaston Bachelard, *The Psychoanalysis of Fire* (Boston: Beacon Press, 1964); Stephen J. Pyne, *Fire: Nature and Culture* (London: Reaktion Books, 2012).

4. David E. Nye, *When the Lights Went Out: A History of Blackouts in America* (Cambridge, MA: MIT Press, 2010), 96.

5. Margaret E. Morris, Douglas M. Carmean, Artem Minyaylov, and Luis Ceze, "Augmenting Interpersonal Communication through Connected Lighting," in *Proceedings of the 2017 CHI Conference Extended Abstracts on Human Factors in Computing Systems* (New York: ACM, 2017), 1924–1931.

6. Michal D. Robinson, L. Elizabeth Crawford, and Whitney J. Ahlvers, "When 'Light' and 'Dark' Thoughts Become Light and Dark Responses: Affect Biases Brightness Judgments," *Emotion* 7, no. 2 (May 2007): 366–376.

7. Lucas A. Keefer, Mark J. Landau, Daniel Sullivan, and Zachary K. Rothschild, "Embodied Metaphor and Abstract Problem Solving: Testing a Metaphoric Fit Hypothesis in the Health Domain," *Journal of Experimental Social Psychology* 55 (2014): 12–20.

8. Jeanne L. Tsai, "Ideal Affect: Cultural Causes and Behavioral Consequences," *Perspectives on Psychological Science* 2, no. 3 (2007): 242–259.

9. Richard P. Sloan, Peter A. Shapiro, J. Thomas Bigger, Emilia Bagiella, Richard C. Steinman, and Jack M. Gorman, "Cardiac Autonomic Control and Hostility in Healthy Subjects," *American Journal of Cardiology* 74, no. 3 (1994): 298–300; Yoichi Chida and Andrew Steptoe, "The Association of Anger and Hostility with Future Coronary Heart Disease: A Meta-Analytic Review of Prospective Evidence," *Journal of the American College of Cardiology* 53, no. 11 (2009): 936–946; Bruce S. McEwen and Elizabeth Norton Lasley, *The End of Stress as We Know It* (Washington, DC: Joseph Henry Press, 2002).

10. Roger Fisher, William L. Ury, and Bruce Patton, *Getting to Yes: Negotiating Agreement without Giving In* (New York: Penguin Books, 2011).

11. Heschong, *Thermal Delight in Architecture*, 23.

12. David Sibley, "The Annual Plumage Cycle of a Male American Goldfinch," *Sibley Guides* (blog), May 1, 2012, http://www.sibleyguides.com/2012/05/the-annual -plumage-cycle-of-a-male-american-goldfinch.

13. Cindy Hsin-Liu Kao, Asta Roseway, Bichlien Nguyen, and Michael Dickey, "Earthtones: Chemical Sensing Powders to Detect and Display Environmental Hazards through Color Variation," in *Proceedings of the 2017 CHI Conference Extended Abstracts on Human Factors in Computing Systems* (New York: ACM, 2017), 872–883.

14. "The HugShirt," *CUTECIRCUIT* (blog), April 10, 2014, https://cutecircuit.com /the-hug-shirt; Hyemin Chung, Chia-Hsun Jackie Lee, and Ted Selker, "Lover's Cups: Drinking Interfaces as New Communication Channels," in *CHI '06 Extended Abstracts on Human Factors in Computing Systems* (New York: ACM, 2006), 375–380; Elizabeth Goodman and Marion Misilim, "The Sensing Bed," in *Proceedings of Ubi-Comp 2003* (London: Springer, 2003). For a discussion of computing to support relationships, see Genevieve Bell, Tim Brooke, Elizabeth Churchill, and Eric Paulos, "Intimate Ubiquitous Computing," in *Proceedings of Ubicomp 2003* (New York: ACM, 2003), 3–6.

15. Meghan Clark and Prabal Dutta, "The Haunted House: Networking Smart Homes to Enable Casual Long-Distance Social Interactions," in *Proceedings of the 2015 International Workshop on Internet of Things toward Applications* (New York: ACM, 2015).

16. Mariam Hassib, Max Pfeiffer, Stefan Schneegass, Michael Rohs, and Florian Alt, "Emotion Actuator: Embodied Emotional Feedback through Electroencephalography and Electrical Muscle Stimulation," in *Proceedings of the 2017 CHI Conference on Human Factors in Computing Systems* (New York: ACM, 2017), 6133–6146.

17. Fisher, Ury, and Patton, *Getting to Yes*.

18. Matthew J. Hertenstein, Dacher Keltner, Betsy App, Brittany A. Bulleit, and Ariane R. Jaskolka, "Touch Communicates Distinct Emotions," *Emotion* 6, no. 3 (2006): 528–533.

19. Guy R. Newsham and Jennifer A. Veitch, "Lighting Quality Recommendations for VDT Offices: A New Method of Derivation," *Transactions of the Illuminating Engineering Society* 33, no. 2 (June 1, 2001): 97–113; Robert Baron, Mark Rea, and Susan G. Daniels, "Effects of Indoor Lighting (Illuminance and Spectral Distribution) on the Performance of Cognitive Tasks and Interpersonal Behaviors: The Potential Mediating Role of Positive Affect," *Motivation and Emotion* 16 (January 3, 1992): 1–33; Richard D. Barnes, "Perceived Freedom and Control in the Built Environment," in *Cognition, Social Behavior, and the Environment*, ed. John H. Harvey (Mahwah, NJ: Lawrence Erlbaum, 1981), 409–422.

20. Albert Bandura, "Self-Efficacy Mechanism in Human Agency," *American Psychologist* 37 (February 1982): 122–147.

21. Lyn Y. Abramson, Martin E. Seligman, and John D. Teasdale, "Learned Helplessness in Humans: Critique and Reformulation," *Journal of Abnormal Psychology* 87, no. 1 (1978): 49–74; Steven F. Maier and Martin E. Seligman, "Learned Helplessness: Theory and Evidence," *Journal of Experimental Psychology: General* 105, no. 1 (1976): 3–46.

22. Described in Nancy K. Baym, *Personal Connections in the Digital Age* (Hoboken, NJ: John Wiley and Sons, 2015), 70.

23. Luke Stark and Kate Crawford, "The Conservatism of Emoji: Work, Affect, and Communication," *Social Media + Society* 1, no. 2 (September 2015): 1–11.

24. Emily T. Amanatullah and Michael W. Morris, "Negotiating Gender Roles: Gender Differences in Assertive Negotiating Are Mediated by Women's Fear of Backlash and Attenuated When Negotiating on Behalf of Others," *Journal of Personality and Social Psychology* 98, no. 2 (2010): 256–267; Hannah Riley Bowles and Linda Babcock, "How Can Women Escape the Compensation Negotiation Dilemma? Relational Accounts Are One Answer," *Psychology of Women Quarterly* 37, no. 1 (2013): 80–96.

25. Bowles and Babcock's research indicates that in addition to combining warmth and dominance, women have more success in negotiations when they present "relational accounts": explanations of how a proposed change serves the interest of an organization.

26. Sheryl Sandberg, *Lean In: Women, Work, and the Will to Lead* (New York: Alfred A. Knopf, 2013), 63.

27. Kate M. Miltner, "'There's No Place for Lulz on LOLCats': The Role of Genre, Gender, and Group Identity in the Interpretation and Enjoyment of an Internet Meme," *First Monday* 19, no. 8 (August 2014), http://firstmonday.org/ojs/index.php /fm/article/view/5391.

28. Bronislaw Malinowski, *The Problem of Meaning in Primitive Languages: Supplement 1* (Abingdon, UK: Routledge and Kegan Paul, 1949).

29. Jonathan Donner, "The Rules of Beeping: Exchanging Messages via Intentional 'Missed Calls' on Mobile Phones," *Journal of Computer-Mediated Communication* 13, no. 1 (October 2007): 1–22.

Chapter 2: Conversational Catalysts

1. Ian Bogost, "Persuasive Games: Words with Friends Forever," *Gamasutra* (blog), October 2, 2012, https://www.gamasutra.com/view/feature/178658/persuasive_games _words_with_.php.

2. Ian Bogost, "Video Games Are Better without Characters," *Atlantic*, March 13, 2015, https://www.theatlantic.com/technology/archive/2015/03/video-games-are -better-without-characters/387556.

3. Theresa M. Senft, *Camgirls: Celebrity and Community in the Age of Social Networks* (Bern, Switzerland: Peter Lang, 2008).

4. Alice E. Marwick, "You May Know Me from YouTube: (Micro-)Celebrity in Social Media," in *A Companion to Celebrity*, ed. P. David Marshall and Sean Redmond (Hoboken, NJ: John Wiley and Sons, 2015), 333–350.

5. R. I. M. Dunbar, "Gossip in Evolutionary Perspective," *Review of General Psychology* 8 (2004): 100–110.

6. In *Status Update: Celebrity, Publicity, and Branding in the Social Media Age* (New Haven, CT: Yale University Press, 2013), Alice E. Marwick points out that microcelebrities lack the professional support and funding to protect themselves from online critique. Transparency and exhibitionism are almost demanded of those building an audience online, and this makes them vulnerable to harsh criticism.

7. For research on conveying authenticity, see Alice E. Marwick and danah boyd, "I Tweet Honestly, I Tweet Passionately: Twitter Users, Context Collapse, and the Imagined Audience," *New Media and Society* 13, no. 1 (2011): 114–133.

8. The Mood Phone was developed as a research prototype. It is not commercially available. For a description of the application and trials, see Margaret E. Morris, Qusai Kathawala, Todd K. Leen, Ethan E. Gorenstein, Farzin Guilak, William DeLeeuw, and Michael Labhard, "Mobile Therapy: Case Study Evaluations of a Cell Phone Application for Emotional Self-Awareness," *Journal of Medical Internet Research* 12, no. 2 (April 2010).

9. Rosalind W. Picard, *Affective Computing* (Cambridge, MA: MIT Press, 2000). The dimensions of arousal and valence are described in the circumplex model: James A. Russell, "A Circumplex Model of Affect," *Journal of Personality and Social Psychology* 39, no. 6 (1980): 1161.

10. Lisa Feldman Barrett, *How Emotions Are Made: The Secret Life of the Brain* (Boston: Houghton Mifflin Harcourt, 2017).

11. Lisa Feldman Barrett, "Are You in Despair? That's Good," *New York Times*, June 3, 2016, Opinion section, https://www.nytimes.com/2016/06/05/opinion/sunday /are-you-in-despair-thats-good.html.

12. A compelling example is WeFeelFine (http://wefeelfine.org), which displays how users around the world are feeling, in their own words.

13. "Boomerang Respondable," https://www.boomeranggmail.com/respondable; "Discover and Share Your Personality," Crystal, https://www.crystalknows.com.

14. Cristian Danescu-Niculescu-Mizil, Moritz Sudhof, Dan Jurafsky, Jure Leskovec, and Christopher Potts, "A Computational Approach to Politeness with Application to Social Factors," in *Proceedings of the 51st Annual Meeting of the Association for Computational Linguistics* (Sofia, Bulgaria: ACL, 2013), 250–259; Moira Burke and Robert Kraut, "Mind Your Ps and Qs: The Impact of Politeness and Rudeness in Online Communities," in *Proceedings of the 2008 ACM Conference on Computer Supported Cooperative Work* (New York: ACM, 2008), 281–284.

15. Molly E. Ireland and James W. Pennebaker, "Language Style Matching in Writing: Synchrony in Essays, Correspondence, and Poetry," *Journal of Personality and Social Psychology* 99, no. 3 (2010): 549–571; James W. Pennebaker, Martha E. Francis, and Roger J. Booth, *Linguistic Inquiry and Word Count: LIWC 2001* (Mahwah, NJ: Lawrence Erlbaum, 2001). The feedback on matching in email was developed in collaboration with a team at Intel, including Doug Carmean, Janet Tseng, Cindy Chung, and Adam Laskowitz, and studied in collaboration with University of Washington researchers Sean Munson, Conrad Nied, and Gary Hsieh.

16. "Tone Analyzer," IBM, Watson, https://www.ibm.com/watson/services/tone -analyzer.

17. For a review, see Dan Jurafsky and James H. Martin, *Speech and Language Processing* (London: Pearson, 2014).

18. See Lauren McCarthy, "Social Turkers: Crowdsourced Dating," http://socialturk ers.com.

19. For details, see "Crowdpilot," http://www.crowdpilot.me. I was a collaborator but have no financial interest in the app.

20. Taeyoon Choi, "Repo for My Dating," December 9, 2017, https://github.com /tchoi8/dating.

21. Lisa Heschong, *Thermal Delight in Architecture* (Cambridge, MA: MIT Press, 1979), 72.

22. This type of sharing also played out among colleagues, couples, and friends in trials of the Mood Phone.

23. Sherry Turkle, *Evocative Objects: Things We Think With* (Cambridge, MA: MIT Press, 2011).

24. Donald Woods Winnicott, "Transitional Objects and Transitional Phenomena: A Study of the First Not-Me Possession," *International Journal of Psychoanalysis* 34, no. 2 (1953): 89–97.

25. Sherry Turkle, *Reclaiming Conversation: The Power of Talk in a Digital Age* (New York: Penguin Books, 2016).

Chapter 3: Meaningful Measures

1. Susannah Fox and Maeve Duggan, "Tracking for Health," Pew Research Center, January 28, 2013, http://www.pewinternet.org/2013/01/28/tracking-for-health.

2. "Seth Roberts on Arithmetic and Butter," Quantified Self Labs, http://quantified self.com/2010/09/seth-roberts-on-arithmetic-and.

3. Gary Wolf, "The Data-Driven Life," *New York Times*, April 28, 2010, http://www .nytimes.com/2010/05/02/magazine/02self-measurement-t.html.

4. "QS Conference 2012–Kevin Kelly–Conference Closing Keynote," 2012, https:// vimeo.com/56082231.

5. "Wearable Sensors Can Tell When You Are Getting Sick," News Center, http:// med.stanford.edu/news/all-news/2017/01/wearable-sensors-can-tell-when-you-are -getting-sick.html.

6. Elizabeth O. Lillie, Bradley Patay, Joel Diamant, Brian Issell, Eric J. Topol, and Nicholas J. Schork, "The N-of-1 Clinical Trial: The Ultimate Strategy for Individualizing Medicine?," *Personalized Medicine* 8, no. 2 (2011): 161–173; Predrag Klasnja, E. B. Hekler, S. Shiffman, A. Boruvka, D. Almirall, A. Tewari, and S. A. Murphy, "Microrandomized Trials: An Experimental Design for Developing Just-in-Time Adaptive Interventions," *Health Psychology* 34, no. S (2015): 1220–1228; Ravi Karkar, Jessica Schroeder, Daniel A. Epstein, Laura R. Pina, Jeffrey Scofield, James Fogarty, Julie A. Kientz, et al., "TummyTrials: A Feasibility Study of Using Self-Experimentation to Detect Individualized Food Triggers," in *Proceedings of the SIGCHI Conference on Human Factors in Computing Systems* (New York: ACM, 2017): 6850–6863. See also, Katie Thomas, "His Doctors Were Stumped. Then He Took Over," *New York Times*, February 4, 2017, https://www.nytimes.com/2017/02/04/business/his-doctors-were -stumped-then-he-took-over.html.

7. Jennifer Mankoff, personal communication; https://gotlyme.wordpress.com.

8. Matthew J. Bietz, Gillian R. Hayes, Margaret E. Morris, Heather Patterson, and Luke Stark, "Creating Meaning in a World of Quantified Selves," *IEEE Pervasive Computing* 15, no. 2 (April 2016): 82–85.

9. Emotional engines were promoted by several car manufacturers at the 2018 Consumer Electronic Show; see Steve Winter and Kenny Fried, "Emotion-Driven Vehicle AI, Driverless Transit among Auto Innovations at CES," WTOP, January 9, 2017, https://wtop.com/tech/2017/01/emotion-fueled-vehicle-ai-driverless -personal-transit-pave-road-future-innovation-ces. Affectiva announced its emotional awareness system for cars; see "Affectiva Automotive AI," Affectiva, https:// www.affectiva.com/product/affectiva-automotive-ai. Research on measuring stress among drivers has a much longer history; see, Jennifer A. Healey and Rosalind W. Picard, "Detecting Stress during Real-World Driving Tasks Using Physiological

Sensors," *Transactions on Intelligent Transportation Systems* 6, no. 2 (June 2005): 156–166.

10. For examples, see http://www.huahuacaocao.com/product; "Home Has a New Hub," Samsung, https://www.samsung.com/us/explore/family-hub-refrigerator/con nected-hub.

11. For a discussion of tensions involving self-tracking data, see Gina Neff and Dawn Nafus, *Self-Tracking* (Cambridge, MA: MIT Press, 2016); Dawn Nafus and Jamie Sherman, "This One Does Not Go up to 11: The Quantified Self Movement as an Alternative Big Data Practice," *International Journal of Communication* 8 (June 2014): 1784–1794.

12. Natasha Dow Schüll, "Data for Life: Wearable Technology and the Design of Self-Care," *BioSocieties* 11, no. 3 (2016): 317–333.

13. Natasha Dow Schüll, "Tracking," in *Experience: Culture, Cognition, and the Common Sense* (Cambridge, MA: MIT Press, 2016), 195–203.

14. Natasha Dow Schüll, "Self in the Loop: Bits, Patterns, and Pathways in the Quantified Self," in *The Networked Self*, vol. 5, *Human Augmentics, Artificial Intelligence, Sentience*, ed. Zizi Papacharisi (Abingdon, UK: Routledge, 2018).

15. For research on designing to motivate behavior change, see Sunny Consolvo, David W. McDonald, and James A. Landay, "Theory-Driven Design Strategies for Technologies That Support Behavior Change in Everyday Life," in *Proceedings of the SIGCHI Conference on Human Factors in Computing Systems* (New York: ACM, 2009), 405–414; Predrag Klasnja, Sunny Consolvo, David W. McDonald, James A. Landay, and Wanda Pratt, "Using Mobile and Personal Sensing Technologies to Support Health Behavior Change in Everyday Life," *AMIA Annual Symposium Proceedings Archive* (November 2009): 338–342; BJ Fogg, "A Behavior Model for Persuasive Design," in *Proceedings of the 4th International Conference on Persuasive Technology* (New York: ACM, 2009), 40:1–7. For critical commentary on persuasive design in the context of mindfulness and productivity, see Melissa Greg, *Counterproductive: Time Management after the Organization* (Durham, NC: Duke University Press, 2018).

16. Kaiton Williams, personal communication; Kaiton Williams, "An Anxious Alliance," in *Proceedings of the Fifth Decennial Aarhus Conference on Critical Alternatives* (Aarhus, Denmark: Aarhus University Press, 2015), 121–131.

17. Chia-Fang Chung, Elena Agapie, Jessica Schroeder, Sonali Mishra, James Fogarty, and Sean A. Munson, "When Personal Tracking Becomes Social: Examining the Use of Instagram for Healthy Eating," *Proceedings of the SIGCHI Conference on Human Factors in Computing Systems* (New York: ACM, 2017): 1674–1687.

18. Stephanie Zerwas, associate professor at the University of North Carolina School of Medicine, in personal communication, explained that apps for diet and movement

tracking can reinforce the obsessive self-monitoring among people with anorexia and trigger symptoms in those who are vulnerable due to conscientiousness and perfectionist tendencies.

19. Daniel Victor, "Google Maps Pulls Calorie-Counting Feature after Criticism," *New York Times*, October 17, 2017, https://www.nytimes.com/2017/10/17/technology /google-maps-calories.html?_r=0.

20. This system was developed in collaboration with Intel colleagues Doug Carmean, Janet Tseng, Cindy Chung, and Adam Laskowitz, and studied in collaboration with University of Washington researchers Sean Munson, Conrad Nied, and Gary Hsieh.

21. Molly E. Ireland and James W. Pennebaker, "Language Style Matching in Writing: Synchrony in Essays, Correspondence, and Poetry," *Journal of Personality and Social Psychology* 99, no. 3 (2010): 549–571; Noah Liebman and Darren Gergle, "It's (Not) Simply a Matter of Time: The Relationship between CMC Cues and Interpersonal Affinity," in *Proceedings of the 19th ACM Conference on Computer-Supported Cooperative Work and Social Computing* (New York: ACM, 2016), 570–581; Roderick I. Swaab, William W. Maddux, and Marwan Sinaceur, "Early Words That Work: When and How Virtual Linguistic Mimicry Facilitates Negotiation Outcomes," *Journal of Experimental Social Psychology* 47, no. 3 (2011): 616–621; William W. Maddux, Elizabeth Mullen, and Adam D. Galinsky, "Chameleons Bake Bigger Pies and Take Bigger Pieces: Strategic Behavioral Mimicry Facilitates Negotiation Outcomes," *Journal of Experimental Social Psychology* 44, no. 2 (2008): 461–468.

22. This research drew on the LIWC program and associated research. See James W. Pennebaker, Martha E. Francis, and Roger J. Booth, *Linguistic Inquiry and Word Count: LIWC 2001* (Mahwah, NJ: Lawrence Erlbaum, 2001); Yla R. Tausczik and James W. Pennebaker, "The Psychological Meaning of Words: LIWC and Computerized Text Analysis Methods," *Journal of Language and Social Psychology* 29, no. 1 (2010): 24–54.

23. Louise C. Hawkley and John T. Cacioppo, "Loneliness Matters: A Theoretical and Empirical Review of Consequences and Mechanisms," *Annals of Behavioral Medicine: A Publication of the Society of Behavioral Medicine* 40, no. 2 (October 2010): 218–227.

24. Margaret E. Morris, "Social Networks as Health Feedback Displays," *IEEE Internet Computing* 9, no. 5 (September 2005): 29–37. This project was conducted with researchers Eric Dishman, Jay Lundell, Brad Needham, and Terry Dishongh.

25. This example was relayed to me by George's friend.

Chapter 4: Remembering and Forgetting

1. Casey Newton, "Speak, Memory: When Her Best Friend Died, She Used Artificial Intelligence to Keep Talking to Him," *Verge*, October 6, 2016, http://www.theverge .com/a/luka-artificial-intelligence-memorial-roman-mazurenko-bot.

2. Chris Foxx, "App Creates Selfies with Dead Relatives," BBC News, March 2, 2017, http://www.bbc.com/news/av/technology-39130792/app-creates-selfies-with -avatars-of-dead-relatives.

3. Jed R. Brubaker, Gillian R. Hayes, and Paul Dourish, "Beyond the Grave: Facebook as a Site for the Expansion of Death and Mourning," *Information Society* 29, no. 3 (2013): 152–163; Alice E. Marwick and Nicole B. Ellison, "There Isn't Wifi in Heaven! Negotiating Visibility on Facebook Memorial Pages," *Journal of Broadcasting and Electronic Media* 3 (2012): 378–400.

4. Brubaker, Hayes, and Dourish, "Beyond the Grave."

5. Kathleen M. Cumiskey and Larissa Hjorth, *Haunting Hands: Mobile Media Practices and Loss* (New York: Oxford University Press, 2017).

6. Marwick and Ellison, "There Isn't Wifi in Heaven!"

7. This example is based on an interview and Dana Greenfield, "Leaning into Grief," Quantified Self, http://quantifiedself.com/2014/05/dana-greenfield-leaning-grief.

8. Nicky Woolf and Olivia Solon, "Facebook Profile Glitch 'Kills' Millions. Even Mark Zuckerberg," *Guardian*, November 11, 2016, Technology section, http://www .theguardian.com/technology/2016/nov/11/facebook-profile-glitch-deaths-mark -zuckerberg.

9. Corina Sas and Steve Whittaker, "Design for Forgetting: Disposing of Digital Possessions after a Breakup," in *Proceedings of the SIGCHI Conference on Human Factors in Computing Systems* (New York: ACM, 2013), 1823–1832.

10. Jed R. Brubaker, Funda Kivran-Swaine, Lee Taber, and Gillian R. Hayes, "Grief-Stricken in a Crowd: The Language of Bereavement and Distress in Social Media," in *Proceedings of the Sixth International AAAI Conference on Weblogs and Social Media* (Palo Alto, CA: AAAI Press, 2012).

11. Lars Schwabe, Karim Nader, and Jens C. Pruessner, "Reconsolidation of Human Memory: Brain Mechanisms and Clinical Relevance," *Biological Psychiatry* 76, no. 4 (August 15, 2014): 274–280; Merel Kindt, Marieke Soeter, and Bram Vervliet, "Beyond Extinction: Erasing Human Fear Responses and Preventing the Return of Fear," *Nature Neuroscience* 12, no. 3 (2009): 256–258.

12. Corina Sas, Steve Whittaker, and John Zimmerman, "Design for Rituals of Letting Go: An Embodiment Perspective on Disposal Practices Informed by Grief Therapy," *ACM Transactions on Computer-Human Interaction* 23, no. 4 (2016): 1–37.

13. This interview was part a research program at Intel on how computing could meet the needs of older adults and their families, particularly those coping with cognitive decline. For a discussion of the barriers to early detection that emerged from ethnographic research, see Margaret Morris, Jay Lundell, Eric Dishman, and Brad Needham, "New Perspectives on Ubiquitous Computing from Ethnographic

Study of Elders with Cognitive Decline," in *International Conference on Ubiquitous Computing* (Berlin: Springer, 2003), 227–242; Margaret Morris, Stephen S. Intille, and Jennifer S. Beaudin, "Embedded Assessment: Overcoming Barriers to Early Detection with Pervasive Computing," in *Proceedings of the Third International Conference on Pervasive Computing* (Berlin: Springer, 2005), 333–346.

14. For a review of temporal discounting, see Shane Frederick, George Loewenstein, and Ted O'Donoghue, "Time Discounting and Time Preference: A Critical Review," *Journal of Economic Literature* 40, no. 2 (2002): 351–401; George Ainslie, *Breakdown of Will* (New York: Cambridge University Press, 2001); Howard Rachlin, *The Science of Self-Control* (Cambridge, MA: Harvard University Press, 2000).

15. "About stickK.com," https://www.stickk.com/aboutus.

16. Richard H. Thaler and Cass R. Sunstein, *Nudge: Improving Decisions about Health, Wealth, and Happiness* (New Haven, CT: Yale University Press, 2008).

17. Ainslie, *Breakdown of Will*, 143.

18. Rachlin, *Science of Self-Control*.

19. For a discussion of policies and other forms of intertemporal bargaining, see Rachlin, *The Science of Self-Control*; Ainslie, *Breakdown of Will*.

20. Dawit Shawel Abebe, Leila Torgersen, Lars Lien, Gertrud S. Hafstad, and Tilmann von Soest, "Predictors of Disordered Eating in Adolescence and Young Adulthood: A Population-Based, Longitudinal Study of Females and Males in Norway," *International Journal of Behavioral Development* 38, no. 2 (2014): 128–138; Martha Peaslee Levine, "Loneliness and Eating Disorders," *Journal of Psychology* 146, no. 1–2 (January 1, 2012): 243–257.

21. Jennifer S. Kirby, Cristin D. Runfola, Melanie S. Fischer, Donald H. Baucom, and Cynthia M. Bulik, "Couple-Based Interventions for Adults with Eating Disorders," *Eating Disorders* 23, no. 4 (2015): 356–365; Jennifer S. Kirby, Melanie S. Fischer, Thomas J. Raney, Donald H. Baucom, and Cynthia M. Bulik, "Couple-Based Interventions in the Treatment of Adult Anorexia Nervosa: A Brief Case Example of UCAN," *Psychotherapy* 53, no. 2 (June 2016): 241–250.

22. For an overview, see Thaler and Sunstein, *Nudge*.

23. Environmental influences on obesity are discussed in Kelly D. Brownell, Rogan Kersh, David S. Ludwig, Robert C. Post, Rebecca M. Puhl, Marlene B. Schwartz, and Walter C. Willett, "Personal Responsibility and Obesity: A Constructive Approach to a Controversial Issue," *Health Affairs* 29, no. 3 (March 2010): 379–387; Reid Ewing, Tom Schmid, Richard Killingsworth, Amy Zlot, and Stephen Raudenbush, "Relationship between Urban Sprawl and Physical Activity, Obesity, and Morbidity," in *Urban Ecology: An International Perspective on the Interaction between Humans and Nature*, ed. John M. Marzluff, Eric Shulenberger, Wilfred Endlicher, Marina Alberti,

Gordon Bradley, Clare Ryan, Ute Simon, et al. (Boston: Springer, 2008), 567–582. The social environment is addressed in addiction treatments such as the Community Reinforcement Approach; see Jane Ellen Smith, Robert J. Meyers, and William R. Miller, "The Community Reinforcement Approach to the Treatment of Substance Use Disorders," supplement, *American Journal on Addictions* 10 (2001): 51–59.

24. Mitchell H. Katz, "Structural Interventions for Addressing Chronic Health Problems," *JAMA* 302, no. 6 (August 2009): 683–685.

25. Bradley M. Appelhans, Simone A. French, Tamara Olinger, Michael Bogucki, and Lisa M. Powell, "Leveraging Delay Discounting for Health: Can Time Delays Influence Food Choice?," *Appetite* 126 (July 2018): 16–25.

26. Wendy Wood, Leona Tam, and Melissa Guerrero Witt, "Changing Circumstances, Disrupting Habits," *Journal of Personality and Social Psychology* 88, no. 6 (June 2005): 918–933.

27. As described in Thaler and Sunstein, *Nudge*, 233, several states allow gamblers to join a list that legally bans them from casinos or collecting winnings.

Chapter 5: Beyond the Hookup

1. Anna Moore, "How Tinder Took Me from Serial Monogamy to Casual Sex," *Guardian*, September 28, 2014, https://www.theguardian.com/lifeandstyle/2014/sep/28/tinder-serial-monogamy-casual-sex.

2. See Kath Albury, Jean Burgess, Ben Light, Kane Race, and Rowan Wilken, "Data Cultures of Mobile Dating and Hook-up Apps: Emerging Issues for Critical Social Science Research," *Big Data and Society* 4, no. 2 (2017): 1–11. For examples of books about success in optimizing online dating, see Amy Webb, *Data, a Love Story: How I Cracked the Online Dating Code to Meet My Match* (New York: Penguin Books, 2013); Chris McKinlay, *Optimal Cupid: Mastering the Hidden Logic of OkCupid* (CreateSpace Independent Publishing Platform, 2014). Albury and her colleagues also note some exceptions to generally benign ways to take control over online dating and point to related research; see Ran Almog and Danny Kaplan, "The Nerd and His Discontent: The Seduction Community and the Logic of the Game as a Geeky Solution to the Challenges of Young Masculinity," *Men and Masculinities* 20, no. 1 (2017): 27–48.

3. Jennifer Crocker and Connie T. Wolfe, "Contingencies of Self-Worth," *Psychological Review* 108, no. 3 (2001): 593–623.

4. William W. Maddux, Elizabeth Mullen, and Adam D. Galinsky, "Chameleons Bake Bigger Pies and Take Bigger Pieces: Strategic Behavioral Mimicry Facilitates Negotiation Outcomes," *Journal of Experimental Social Psychology* 44, no. 2 (2008): 461–468.

5. Molly E. Ireland, Richard B. Slatcher, Paul W. Eastwick, Lauren E. Scissors, Eli J. Finkel, and James W. Pennebaker, "Language Style Matching Predicts Relationship

Initiation and Stability," *Psychological Science* 22, no. 1 (January 2011): 39–44; Maddux, Mullen, and Galinsky, "Chameleons Bake Bigger Pies."

6. "About," Bye Felipe, https://www.byefelipe.com/about.

7. See Christian Rudder, *Dataclysm: Love, Sex, Race, and Identity—What Our Online Lives Tell Us about Our Offline Selves* (New York: Broadway Books, 2015), for data on harassment in online dating.

8. This blog entry of *Tart Response* was examined by Kane Race: "Speculative Pragmatism and Intimate Arrangements: Online Hook-Up Devices in Gay Life," *Culture, Health, Sexuality* 17 (July 2014): 1–16. My description is based on Race's article and my interview with the writer of *Tart Response*.

9. Aaron Smith, "15% of American Adults Have Used Online Dating Sites or Mobile Dating Apps," Pew Research Center, February 11, 2016, http://www.pewinternet.org /2016/02/11/15-percent-of-american-adults-have-used-online-dating-sites-or-mobile -dating-apps.

10. John T. Cacioppo, Stephanie Cacioppo, Gian Gonzaga, Elizabeth L. Ogburn, and Tyler J. VanderWeele, "Marital Satisfaction and Break-Ups Differ across On-Line and Off-Line Meeting Venues," *National Academy of Sciences* 10 (2013): 10135–10140.

11. M. J. Rosenfeld, "Marriage, Choice, and Couplehood in the Age of the Internet," *Sociological Science* 4 (2017): 490–510.

12. Cacioppo, Cacioppo, Gonzaga, Ogburn, and VanderWeele, "Marital Satisfaction and Break-Ups."

13. Josue Ortega and Philipp Hergovich, "The Strength of Absent Ties: Social Integration via Online Dating," September 29, 2017, http://arxiv.org/abs/1709.10478.

14. Rudder, *Dataclysm*, 2014; "Race and Attraction, 2009–2014," *OkCupid* (blog), September 10, 2014, https://theblog.okcupid.com/race-and-attraction-2009-2014-107 dcbb4f060.

15. "'Least Desirable'? How Racial Discrimination Plays Out in Online Dating," NPR, Morning Edition, January 9, 2018, https://www.npr.org/2018/01/09/575352051/least -desirable-how-racial-discrimination-plays-out-in-online-dating.

Chapter 6: Picturing Ourselves

1. Kenneth J. Gergen, *The Saturated Self: Dilemmas of Identity in Contemporary Life* (New York: Basic Books, 1991).

2. Sherry Turkle, *Life on the Screen: Identity in the Age of the Internet* (New York: Simon and Schuster, 2011).

3. Dan P. McAdams, *The Stories We Live By: Personal Myths and the Making of the Self* (New York: Guilford Press, 1993).

4. McAdams, *The Stories We Live By*; Roy Schafer, *Retelling a Life: Narration and Dialogue in Psychoanalysis* (New York: Basic Books, 1992).

5. Samuel D. Gosling, Sei Jin Ko, Thomas Mannarelli, and Margaret E. Morris, "A Room with a Cue: Personality Judgments Based on Offices and Bedrooms," *Journal of Personality and Social Psychology* 82, no. 3 (March 2002): 379–398; Samuel D. Gosling, *Snoop: What Your Stuff Says about You* (New York: Basic Books, 2009).

6. Donald Woods Winnicott, "Transitional Objects and Transitional Phenomena: A Study of the First Not-Me Possession," *International Journal of Psychoanalysis* 34, no. 2 (1953): 89–97. For more on the significance of transitional objects throughout life, see Sherry Turkle, *Evocative Objects: Things We Think With* (Cambridge, MA: MIT Press, 2011).

7. For more on attachment theory, see Mary D. Salter Ainsworth, "Attachments beyond Infancy," *American Psychologist* 44, no. 4 (April 1989): 709–716; John Bowlby, *A Secure Base: Clinical Applications of Attachment Theory* (London: Psychology Press, 1988). For a self-help book on this topic, see Amir Levine and Rachel Heller, *Attached: The New Science of Adult Attachment and How It Can Help You Find—and Keep—Love* (New York: Penguin Books, 2010).

8. Sherry Turkle, *The Second Self: Computers and the Human Spirit* (Cambridge, MA: MIT Press, 2005), 5.

9. This system was developed in collaboration with Intel colleagues Doug Carmean, Janet Tseng, Cindy Chung, and Adam Laskowitz, and studied in collaboration with University of Washington researchers Sean Munson, Conrad Nied, and Gary Hsieh.

10. For a description of this system, see Margaret E. Morris, Carl S. Marshall, Mira Calix, Murad Al Haj, James S. MacDougall, and Douglas M. Carmean, "PIXEE: Pictures, Interaction, and Emotional Expression," in *CHI '13 Extended Abstracts on Human Factors in Computing Systems* (New York: ACM, 2013), 2277–2286. The art events where it was shown were part of the Creators Project, ran by Vice Media and Intel.

11. James Meese, Martin Gibbs, Marcus Carter, Michael Arnold, Bjorn Nansen, and Tamara Kohn, "Selfies at Funerals: Mourning and Presencing on Social Media Platforms," *International Journal of Communication* 9 (2015): 1818–1831.

12. Jill W. Rettberg, *Seeing Ourselves through Technology: How We Use Selfies, Blogs, and Wearable Devices to See and Shape Ourselves* (Basingstoke, UK: Palgrave Macmillan, 2014).

13. Lisa Feldman Barrett, *How Emotions Are Made: The Secret Life of the Brain* (Boston: Houghton Mifflin Harcourt, 2017); Michele M. Tugade, Barbara L. Fredrickson, and

Lisa Feldman Barrett, "Psychological Resilience and Positive Emotional Granularity: Examining the Benefits of Positive Emotions on Coping and Health," *Journal of Personality* 72, no. 6 (December 2004): 1161–1190.

14. For a review of strategies used to manage gender transitions on Facebook, see Oliver L. Haimson Jed R. Brubaker, Lynn Dombrowski, and Gillian R. Hayes, "Disclosure, Stress, and Support during Gender Transition on Facebook," *18th ACM Conference on Computer Supported Cooperative Work and Social Computing* (New York: ACM, 2015), 1176–1190.

15. Hazel Markus and Paula Nurius, "Possible Selves," *American Psychologist* 41, no. 9 (1986): 954–969.

16. For a history of this term, see danah boyd, "How 'Context Collapse' Was Coined: My Recollection," *apophenia* (blog), http://www.zephoria.org/thoughts/archives/2013 /12/08/coining-context-collapse.html.

17. danah boyd, *It's Complicated: The Social Lives of Networked Teens* (New Haven, CT: Yale University Press, 2014).

18. Alice E. Marwick and danah boyd, "Networked Privacy: How Teenagers Negotiate Context in Social Media," *New Media and Society* 16, no. 7 (November 2014): 1051–1067.

Chapter 7: Micro Connections

1. Noam Scheiber, "How Uber Uses Psychological Tricks to Push Its Drivers' Buttons," *New York Times*, April 2, 2017, Technology section, https://www.nytimes.com /interactive/2017/04/02/technology/uber-drivers-psychological-tricks.html.

2. "Introduction," Airbnb, https://blog.atairbnb.com/belong-anywhere.

3. Dan P. McAdams, *The Stories We Live By: Personal Myths and the Making of the Self* (New York: Guilford Press, 1993).

4. McAdams, *The Stories We Live By.*

5. "Why Some Social Network Services Work and Others Don't—Or: the Case for Object-Centered Sociality, *Zengestrom* (blog), April 13, 2005, http://www.zengestrom .com/blog/2005/04/why-some-social-network-services-work-and-others-dont-or-the -case-for-object-centered-sociality.html.

6. Tammy English and Laura L. Carstensen, "Selective Narrowing of Social Networks across Adulthood Is Associated with Improved Emotional Experience in Daily Life," *International Journal of Behavioral Development* 38, no. 2 (March 2014): 195–202.

7. Zheping Huang, "China's Lonely Upper Middle Class Drive Uber Looking for Friends," *Quartz* (blog), https://qz.com/560329/for-chinas-upper-middle-class-driving -for-uber-is-a-cure-for-loneliness.

8. For research on how people form impressions about personality from personal environments, see Samuel D. Gosling, Sei Jin Ko, Thomas Mannarelli, and Margaret E. Morris, "A Room with a Cue: Personality Judgments Based on Offices and Bedrooms," *Journal of Personality and Social Psychology* 82, no. 3 (March 2002): 379–398.

9. John T. Cacioppo and William Patrick, *Loneliness: Human Nature and the Need for Social Connection* (New York: W. W. Norton and Company, 2008).

Chapter 8: Therapy, Virtually

1. "Crisis Trends," Crisis Text Line, https://crisistrends.org.

2. Adam O. Horvath and Lester Luborsky, "The Role of the Therapeutic Alliance in Psychotherapy," *Journal of Consulting and Clinical Psychology* 61, no. 4 (1993): 561–573.

3. Tim Althoff, Kevin Clark, and Jure Leskovec, "Large-Scale Analysis of Counseling Conversations: An Application of Natural Language Processing to Mental Health," *Transactions of the Association for Computational Linguistics* 4 (2016): 463–476.

4. For research on peer support chats for mental health, see Katie O'Leary, Stephen Schueller, Jake Wobbrock, and Wanda Pratt, "Suddenly, We Got to Become Therapists for Each Other": Designing Peer Support Chats for Mental Health," in *Proceedings of the 2018 CHI Conference on Human Factors in Computing Systems* (New York: ACM, 2018), 1–14.

5. This account is based on personal communication with Ronni Higger and this profile: "Can Texting Save Lives?," *New York Times*, https://www.nytimes.com/video /opinion/100000005178504/can-texting-save-lives.html.

6. Pinar Ozturk, Nazanin Andalibi, and Andrea Forte, "Sensitive Self-Disclosures, Responses, and Social Support on Instagram: The Case of #Depression," in *Proceedings of the 2017 ACM Conference on Computer Supported Cooperative Work and Social Computing* (New York: ACM, 2017), 1485–1500; Nazanin Andalibi, Oliver L. Haimson, Munmun De Choudhury, and Andrea Forte, "Understanding Social Media Disclosures of Sexual Abuse through the Lenses of Support Seeking and Anonymity," in *Proceedings of the 2016 CHI Conference on Human Factors in Computing Systems* (New York: ACM, 2016), 3906–3918.

7. Ozturk, Andalibi, and Forte, "Sensitive Self-Disclosures."

8. Andalibi et al., "Understanding Social Media Disclosures."

9. Ozturk, Andalibi, and Forte, "Sensitive Self-Disclosures."

10. Andalibi et al., "Understanding Social Media Disclosures."

11. Nazanin Andalibi, "Self-Disclosure and Response Behaviors in Socially Stigmatized Contexts on Social Media" (PhD diss., Drexel University, 2018).

12. Lydia Manikonda and Munmun De Choudhury, "Modeling and Understanding Visual Attributes of Mental Health Disclosures in Social Media," in *Proceedings of the 2017 CHI Conference on Human Factors in Computing Systems* (New York: ACM, 2017), 170–181.

13. Ozturk, Andalibi, and Forte, "Sensitive Self-Disclosures."

14. Joseph Weizenbaum, "ELIZA—A Computer Program for the Study of Natural Language Communication between Man and Machine," *Communications of the ACM 9*, no. 1 (January 1966): 36–45.

15. Joseph Weizenbaum, "Computer Power and Human Reason," *ANNALS of the American Academy of Political and Social Science 426*, no. 1 (July 1976): 266–267.

16. Sherry Turkle, *The Second Self: Computers and the Human Spirit* (Cambridge, MA: MIT Press, 2005), 42. For further discussion of ELIZA, see Sherry Turkle, "Authenticity in the Age of Digital Companions," *Interaction Studies—Social Behavior and Communication in Biological and Artificial Systems 8*, no. 3 (January 2007): 501–517.

17. Laura Stevens, "'Alexa, Can You Prevent Suicide?' How Amazon Trains Its AI to Handle the Most Personal Questions Imaginable," *Wall Street Journal*, October 23, 2017, Life section, https://www.wsj.com/articles/alexa-can-you-prevent-suicide-1508762311.

18. Woebot, https://www.woebot.io.

19. Kathleen Kara Fitzpatrick, Alison Darcy, and Molly Vierhile, "Delivering Cognitive Behavior Therapy to Young Adults with Symptoms of Depression and Anxiety Using a Fully Automated Conversational Agent (Woebot): A Randomized Controlled Trial," *JMIR Mental Health 4*, no. 2 (2017).

20. Quoted in Fitzpatrick, Darcy, and Vierhile, "Delivering Cognitive Behavior Therapy to Young Adults."

21. Stanford researcher David Lim and Woebot founder Alison Darcy shared these findings in personal communication. Lim provided the caveat that since these individuals completed the trial of Woebot, they may have more positive feelings about it than would be found in the general population.

22. Gale M. Lucas, Jonathan Gratch, Aisha King, and Louis-Philippe Morency, "It's Only a Computer: Virtual Humans Increase Willingness to Disclose," *Computers in Human Behavior 37* (August 2014): 94–100.

23. Xia Jia, "The Psychology Game," *Clarkesworld*, http://clarkesworldmagazine.com /xia_10_17.

24. Nathan Grayson, "Pokémon Go Helped Me Cope with My Social Anxiety," Kotaku, https://kotaku.com/pokemon-go-helped-me-cope-with-my-social-anxiety -1783988220. Researchers have examined the social benefits of Pokémon Go to drive discussions about designing games for social anxiety. See, for example, Adri

Khalis and Amori Yee Mikami, "Who's Gotta Catch 'Em All?: Individual Differences in Pokémon Go Gameplay Behaviors," *Personality and Individual Differences* 124 (April 2018): 35–38.

25. Chris Milk, "How Virtual Reality Can Create the Ultimate Empathy Machine," TED2015, https://www.ted.com/talks/chris_milk_how_virtual_reality_can_create _the_ultimate_empathy_machine.

26. Jaron Lanier, *Dawn of the New Everything: A Journey through Virtual Reality* (New York: Random House, 2017).

27. Jeremy Bailenson, "Empathy/Diversity," VHIL, https://vhil.stanford.edu/empathy-diversity; Hunter G. Hoffman, Gloria T. Chambers, Walter J. Meyer III, Lisa L. Arceneaux, William J. Russell, Eric J. Seibel, Todd L. Richards, et al., "Virtual Reality as an Adjunctive Non-Pharmacologic Analgesic for Acute Burn Pain during Medical Procedures," *Annals of Behavioral Medicine: A Publication of the Society of Behavioral Medicine* 41, no. 2 (April 2011): 183–191;Robert N. McLay, Kenneth Graap, James Spira, Karen Perlman, Scott Johnston, Barbara O. Rothbaum, JoAnn Difede et al., "Development and Testing of Virtual Reality Exposure Therapy for Post-Traumatic Stress Disorder in Active Duty Service Members Who Served in Iraq and Afghanistan," *Military Medicine* 177, no. 6 (June 2012): 635–642; Thomas D. Parsons and Albert A. Rizzo, "Affective Outcomes of Virtual Reality Exposure Therapy for Anxiety and Specific Phobias: A Meta-Analysis," *Journal of Behavior Therapy and Experimental Psychiatry* 39, no. 3 (September 2008): 250–261; "Virtual Reality Clinical Research Laboratory," University of Houston, Graduate School of Social Work, https://ssl.uh .edu/socialwork/New_research/VRCRL/index.

28. Parsons and Rizzo, "Affective Outcomes."

29. For an example of clinically administered virtual reality therapy, see "Virtual Reality for Mental Health," Limbix, https://www.limbix.com/. For a discussion of its use, see Cade Metz, "A New Way for Therapists to Get Inside Heads: Virtual Reality," *New York Times*, July 30, 2017, Technology section, https://www.nytimes.com/2017 /07/30/technology/virtual-reality-limbix-mental-health.html.

30. Richard J. Davidson, Jon Kabat-Zinn, Jessica Schmacher, Melissa Rosenkranz, Daniel Muller, Saki F. Santorelli, Ferris Urbanowski, et al., "Alterations in Brain and Immune Function Produced by Mindfulness Meditation," *Psychosomatic Medicine* 65, no. 4 (2003): 564–570; Madhav Goyal, Sonal Singh, Erica Sibinga, Neda F. Gould, Anastasia Rowland-Seymour, Ritu Sharma, Zackary Berger, et al., "Meditation Programs for Psychological Stress and Well-Being: A Systematic Review and Meta-Analysis," *JAMA Internal Medicine* 174, no. 3 (March 2014): 357–368; Daniel J. Siegel, *Mindsight: The New Science of Personal Transformation* (New York: Bantam Books, 2009).

31. Chip Heath and Dan Heath, *Decisive: How to Make Better Choices in Life and Work* (New York: Crown Business, 2013).

32. Sofia Adelaide Osimo, Rodrigo Pizzaro, Bernhard Spanlang, and Mel Slater, "Conversations between Self and Self as Sigmund Freud—A Virtual Body Ownership Paradigm for Self Counselling," *Scientific Reports* 5 (September 2015): 13899.

33. Jean-Pascal Lefaucheur, Andrea Antal, Samar S. Ayache, David H. Benninger, Jérôme Brunelin, Filippo Cogiamanian, Maria Cotelli, et al., "Evidence-Based Guidelines on the Therapeutic Use of Transcranial Direct Current Stimulation (tDCS)," *Clinical Neurophysiology* 128, no. 1 (January 2017): 56–92; Brian A. Coffman, Vincent P. Clark, and Raja Parasuraman, "Battery Powered Thought: Enhancement of Attention, Learning, and Memory in Healthy Adults Using Transcranial Direct Current Stimulation," *NeuroImage* 85, no. 3 (January 2014): 895–908; Ethan R. Buch, "Effects of TDCS on Motor Learning and Memory Formation: A Consensus and Critical Position Paper," *Clinical Neurophysiology* 128, no. 4 (2017): 589–603.

34. Andre Russowsky Brunoni, Leandro Valiengo, Alessandra Baccaro, Tamires Araujo Zanao, Janaina Farias de Oliveira, Giselly Pereira Vieira, Viviane Freire Bueno, et al., "Sertraline vs Electrical Current Therapy for Treating Depression: Clinical Study: Results from a Factorial, Randomized, Controlled Trial," *JAMA Psychiatry* 70, no. 4 (April 2013): 383–391.

35. Anna Wexler, "Who Uses Direct-to-Consumer Brain Stimulation Products, and Why? A Study of Home Users of TDCS Devices," *Journal of Cognitive Enhancement*, December 27, 2017, 1–21.

36. Wexler notes that these schematics started appearing around 2011 on the tDCS Reddit forum. Wexler, "Who Uses Direct-to-Consumer Brain Stimulation Products, and Why?"

37. Anna Wexler, "The Practices of Do-It-Yourself Brain Stimulation: Implications for Ethical Considerations and Regulatory Proposals," *Journal of Medical Ethics* 42, no. 4 (2016): 211–215.

38. Chris Lp, "Zap Your Brain for a Better You," Engadget, January 13, 2018, https://www.engadget.com/2018/01/13/neurostimulation-tdcs-ces.

39. Rachel Wurzman, Roy H. Hamilton, Alvaro Pascual-Leone, and Michael D. Fox, "An Open Letter concerning Do-It-Yourself Users of Transcranial Direct Current Stimulation," *Annals of Neurology* 80, no. 1 (July 2016): 1–4.

40. Wexler, "Who Uses Direct-to-Consumer Brain Stimulation Products, and Why?" For a discussion of the marketing products (for enhancement versus treatment) to avoid FDA medical device regulation, see Anna Wexler, "Dialing Up Your Brainpower: How Should We Regulate Those Mind-Zapping Gadgets That Promise to Boost Your Memory?," *Slate*, December 18, 2015, http://www.slate.com/articles/technology/future_tense/2015/12/how_should_we_regulate_tdcs_gadgets_that_promise_to_boost_your_brainpower.html.

41. "About Us," Openwater, https://www.openwater.cc/about-us.

42. Jepsen draws on research from Professor Jack Gallant's lab at the University of California at Berkeley on identifying thoughts and mental images through an MRI. See "Brain Viewer," Gallant Lab at UC Berkeley, http://gallantlab.org/index.php /brain-viewer.

43. Mary Lou Jepsen, "Capturing Our Imagination: The Evolution of Brain-Machine Interfaces," MIT Technology Review, November 8, 2017, http://events .technologyreview.com/video/watch/mary-lou-jepsen-openwater-evolution-of -interfaces.

44. The Cleveland Clinic and Case Western Reserve University partnered to create the digital medical school.

45. For a discussion of the modeling technique by Professor Mark Griswold, who developed it, see Microsoft HoloLens, "Microsoft HoloLens: Partner Spotlight with Case Western Reserve University," https://www.youtube.com/watch?v=SKpKlh1 -en0.

46. Although there are tools for the 3-D reconstruction of an image, they do not permit intuitive viewing or navigation of those models. Craig Mundie, personal communication.

Index

Printed in the United States
by Baker & Taylor Publisher Services